The
Grieving
Brain

The
Grieving
Brain

THE SURPRISING SCIENCE OF
HOW WE LEARN FROM
LOVE AND LOSS

Mary-Frances O'Connor

HarperOne
An Imprint of HarperCollins*Publishers*

This book should not be used as a substitute for the advice of a health professional. To protect their privacy, some names have been changed, and in some cases, I have created composites of individuals.

The information in this book has been carefully researched, and all efforts have been made to ensure accuracy. The authors and the publisher assume no responsibility for any injuries suffered or damages or losses incurred during or as a result of following this information. All information should be carefully studied and clearly understood before taking any action based on the information or advice in this book.

Library of Congress Cataloging-in-Publication Data

Names: O'Connor, Mary-Frances, author.
Title: The grieving brain : the surprising science of how we learn from love and
    loss / Mary-Frances O'Connor.
Description: First edition. | New York, NY : HarperOne, [2022]
Identifiers: LCCN 2021022623 | ISBN 9780062946232 (hardcover) | ISBN
    9780062946256 (ebook)
Subjects: LCSH: Grief. | Bereavement—Psychological aspects. | Neuropsychology.
Classification: LCC BF575.G7 O2584 2022 | DDC 155.9/37—dc23
LC record available at https://lccn.loc.gov/2021022623

22 23 24 25 26   LSC   10 9 8 7 6 5 4 3 2 1

To Anna,
for teaching me there is more to life than grief

Contents

Introduction

For as long as there have been human relationships, we have struggled with the overwhelming nature of grief following the death of a loved one. Poets, writers, and artists have given us moving renderings of the almost indescribable nature of loss, an amputation of a part of ourselves, or an absence that weighs on us like a heavy cloak. As human beings, we seem compelled to try to communicate what our grief is like, to describe carrying this burden. In the twentieth century, psychiatrists (Sigmund Freud, Elisabeth Kübler-Ross, and others) began describing from a more objective perspective what people they interviewed felt during grief, and they noted significant patterns and similarities among people. Great descriptions were written into the scientific literature about the "what" of grief—what it feels like, what problems it causes, even what the bodily reactions are.

I always wanted to understand the *why* rather than just the *what*. Why does grief hurt so much? Why does the death, the permanent absence of this person with whom you are bonded, result in such devastating feelings, and lead to behavior and beliefs that are inexplicable, even to you? I felt certain that part of the answer could be found in the brain, the seat of our thoughts and feelings, motivations, and behaviors. If we could look at it from the perspective of what the brain is doing during grief, perhaps we could find the *how*, and that would help us understand the *why*.

People often ask what motivated me to study grief and to become a grief researcher. I think this question usually stems from simple curiosity, but also perhaps from wanting to know if they can trust me. You, too, reading this may want to know if I have walked the walk, through the dark night of death and loss, whether I know of what I speak and study. The grief I have experienced has not been any worse than the grief of others I talk to, who describe their loss and their shattered life in the wake of that loss. But I have known loss. When I was in eighth grade, my mother was diagnosed with stage IV breast cancer. Cancerous cells were in every lymph node the surgeon cut out when he did her mastectomy, so he knew it had traveled already to other parts of her body. Because I was only thirteen, I did not learn until many years later that she was only supposed to survive the year. But I knew that grief came to our house, disrupting our family life, which was already struggling through my parents' separation and my mother's depression. That house sat high in the northern Rocky Mountains near the Continental Divide, in a rural town that benefited from the presence of a small college, where my father taught. My mother's oncologist described her as his "first miracle": she lived another thirteen years—a reprieve from the universe for her two teenage daughters (my older sister and me). But in this world, I was my mother's emotional tonic, her mood regulator. My departure for college, while developmentally appropriate, only compounded her depression. Thus, my desire to understand grief originated not so much from the experience I faced personally after her death, when I was twenty-six, but from a desire to understand my mother's grief and pain in retrospect, and to learn what I could have done to help her.

I went to Northwestern University just outside Chicago and was eager to escape rural life, to be going to college in a city where more people worked in one block than lived in my hometown. The first time I ever came across a mention of functional neuroimaging was when I read a few sentences in my *Introduction to Neuroscience* textbook in the early 1990s. Functional magnetic resonance imaging (fMRI) was brand-new technology, available to only a handful of researchers around the world. I was absolutely intrigued. Although I did not imagine that I would ever have access to such a machine, I was excited about the possibility that scientists could see into the black box of the brain.

Ten years later, in graduate school at the University of Arizona, I completed my dissertation, a study of an intervention for grief. A member of my dissertation committee, a psychiatrist, suggested that I had a great opportunity to see what grief looked like in the brain, and he recommended that I invite the participants from my dissertation study to come back for an fMRI scan. I felt torn. I had already completed the requirements for my PhD in clinical psychology. Neuroimaging was a whole new technology to learn, with quite a steep learning curve. But sometimes all the stars align for a project, and so we began the first fMRI study of grief. The psychiatrist, Richard Lane, had done a sabbatical at University College London, where some of the first methods were developed to analyze images from an fMRI scan. Lane was willing to teach me the analysis, but it still felt like an insurmountable task.

And yet the stars were aligned. It just so happened that a German psychiatrist, Harald Gündel, wanted to come to the United States so Lane could teach him neuroimaging methods, too. Gündel and I first met in March 2000 and felt an immediate connection. We shared a fascination for how the brain maintains the human relationships that help us regulate our emotions, and curiosity about what happens when those relationships are lost. Who would have thought that two researchers, born in two different countries and a decade apart, would overlap so much in their interests? Thus, the pieces of the study were in place. From my dissertation, I knew a group of bereaved people who were willing to undergo a scan. Gündel had the knowledge of brain structure and function. Lane had the imaging skills.

One more obstacle required benevolent intervention from the stars. Gündel was only able to come to the United States for a month; I was heading to UCLA for my clinical internship in July 2001. Alarmingly, the neuroimaging scanner at our university medical center was due to be replaced at the only time when we could all converge in Tucson, Arizona. But all construction projects suffer from the same issue: running behind schedule. So, in May 2001, the scanner schedule was empty, but the older scanner was still available. The first neuroimaging study of grief[1] was done in four weeks, record time for the completion of any research project.

This book offers you the results of that study, in addition to many more.

Moving to UCLA offered me the opportunity to add another area of expertise to my scientific toolkit. I completed my clinical internship there, a year of clinical work in the hospital and clinics, where I saw clients with a range of mental health and medical issues. Following my clinical internship, I embarked on a postdoctoral fellowship in psychoneuroimmunology (PNI), which is a fancy term for studying how immunology fits into our understanding of psychology and neuroscience. I remained at UCLA for ten years, transitioning to faculty, but eventually moved back to the University of Arizona. I run the Grief, Loss and Social Stress (GLASS) lab there, a fulfilling role that enables me to teach graduate and undergraduate students and direct the clinical training program. Now my days are quite varied. I spend hours reading research studies and designing new ones that will probe the mechanisms of the ephemeral experience of grief; I teach undergraduates in classes large and small; I work with other clinical psychologists around the country and the world to help shape the direction of the field of grief research; I mentor graduate students, helping them to develop their own scientific models, write manuscripts to disseminate their findings for the field, and give talks in our local community; and perhaps most important of all, I encourage each student's gift for scientific thinking and urge them to show us their unique view of the world through a scientific lens.

Although my roles as researcher, mentor, professor, and writer mean I no longer see clients in therapy, I have many opportunities to hear about people's grief through extensive interviews I conduct in my research. I ask all sorts of questions, and I also try to listen closely to the kind and generous people who are willing to share their stories with me. Their motivation for participating, they tell me, is to share their experiences with science so that science might be able to help the next person who goes through the awful aftermath of losing a loved one. I am grateful to each of them and have tried to honor their contributions through this book.

Neuroscience is not necessarily the discipline that springs to mind when thinking of grief, and certainly, when my quest began, that was even less the case. Through my years of study and research, I eventually realized the brain has a problem to solve when a loved one has died. This is not a trivial problem. Losing our one-and-only overwhelms us, because we need our loved ones as much as we need food and water.

Fortunately, the brain is good at solving problems. In fact, the brain exists for precisely this function. After decades of research, I realized that the brain devotes lots of effort to mapping where our loved ones are while they are alive, so that we can find them when we need them. And the brain often prefers habits and predictions over new information. But it struggles to learn new information that cannot be ignored, like the absence of our loved one. Grieving requires the difficult task of throwing out the map we have used to navigate our lives together and transforming our relationship with this person who has died. Grieving, or learning to live a meaningful life without our loved one, is ultimately a type of learning. Because learning is something we do our whole lives, seeing grieving as a type of learning may make it feel more familiar and understandable and give us the patience to allow this remarkable process to unfold.

When I talk to students or clinicians or even people sitting next to me on a plane, I find they have burning questions about grief. They ask: Is grief the same as depression? When people do not show their grief, is it because they are in denial? Is losing a child worse than losing a spouse? Then, very often, they ask me this type of question: I know someone whose mom/brother/best friend/husband died, and after six weeks/four months/eighteen months/ten years, they still feel grief. Is this normal?

After many years, it dawned on me that the assumptions behind people's questions demonstrate that grief researchers have not been very successful at broadcasting what they have learned. That is what motivated me to write this book. I am steeped in what psychologist and grief researcher George Bonanno termed *the new science of bereavement*.[2] The type of grief that I focus on in this book applies to those who have lost a spouse, a child, a best friend, or anyone with whom they are close. I also explore other losses, such as the loss of a job, or the pain we feel when a celebrity whom we admire greatly and have never met dies. I offer thoughts for those of us who are adjacent to someone who is grieving, to help us understand what is happening for them. This is not a book of practical advice, and yet many who have read it tell me they learned things they can apply to their own unique experience of loss.

The brain has always fascinated humanity, but new methods allow us to look inside that black box, and what we can see tantalizes us with possible answers to ancient questions. Having said that, I do not believe that a neuroscientific perspective on grief is any better than a sociological, a religious, or an anthropological one. I say that genuinely, despite devoting an entire career to the neurobiological lens. I believe our understanding of grief through a neurobiological lens can enhance our understanding, create a more holistic view of grief, and help us engage in new ways with the anguish and terror of what grief is like. Neuroscience is part of the conversation of our times. By understanding the myriad aspects of grief, by focusing in greater detail on how brain circuits, neurotransmitters, behaviors, and emotions are engaged during bereavement, we have an opportunity to empathize in a new way with those who are currently suffering. We can allow ourselves to feel grief, allow others to feel grief, and understand the experience of grieving—all with greater compassion and hope.

You may have noticed that I use the terms *grief* and *grieving*. Although you hear them used interchangeably, I make an important distinction between them. On the one hand, there is *grief*—the intense emotion that crashes over you like a wave, completely overwhelming, unable to be ignored. Grief is a moment that recurs over and over. However, these moments are distinct from what I call *grieving*, the word I use to refer to the process, not the moment of grief. Grieving has a trajectory. Obviously, grief and grieving are related, which is why the two terms have been used interchangeably when describing our experience of loss. But there are key differences. You see, grief never ends, and it is a natural response to loss. You will experience pangs of grief over this specific person forever. You will have discrete moments that overwhelm you, even years after the death when you have restored your life to a meaningful, fulfilling experience. But, whereas you will feel the universally human emotion of grief forever, your grieving, your adaptation, changes the experience over time. The first one hundred times you have a wave of grief, you may think, "I will never get through this, I cannot bear this." The one hundred and first time, you may think, "I hate this, I don't want this—but it is familiar, and I know I will get through this moment." Even if the feeling of grief is the same, your relationship to the feeling changes. Feeling grief years after your loss may make you doubt whether you have really adapted. If you think of the emotion and the process of adaptation as two different things, however, then it isn't a problem that you experience *grief* even when you have been *grieving* for a long time.

You can think about our journey together through this book as a series of mysteries that we are solving, with part I organized around grief and part II organized around grieving. Each chapter tackles a particular question. Chapter 1 asks, Why is it so hard to understand that the person has died and is gone forever? Cognitive neuroscience helps me to address this question. Chapter 2 asks, Why does grief cause so many emotions—why do we feel such strong sadness, anger, blame, guilt, and yearning? Here I bring in the theory of attachment, including our neural attachment system. Chapter 3 builds on the answers in the first two chapters with a follow-up question: Why does it take so long to understand that our loved one is gone forever? I explain the multiple forms of knowledge that our brain holds simultaneously to think through this puzzle. By chapter 4, we have enough background to dig into a primary question: What happens in the brain during grief? However, to understand the answer to this question we also consider: How has our understanding of grief changed over the history of bereavement science? Chapter 5 looks with more nuance at why some people adapt better than others when they lose a loved one, and asks, What are the complications in complicated grief? Chapter 6 reflects on why it hurts so much when we lose this specific beloved person. This is a chapter about how love works, and how our brain enables the bonding that happens in relationships. Chapter 7 addresses what we can do when we are overwhelmed with grief. I rely on clinical psychology to delve into answers to this question.

In part II we turn to the topic of grieving, and how we might go about restoring a meaningful life. Chapter 8 asks, Why do we ruminate so much after we lose a loved one? Changing what we spend our time thinking about can change our neural connections and increase our chances of learning to live a meaningful life. Turning from focusing on the past, however, leads us to the question in chapter 9, Why would we engage in our life in the present moment, if it is full of grief? The response includes the idea that only in the present moment can we also experience joy and common humanity, and express love to our living loved ones. From the past and present, in chapter 10 we turn to the future and ask, How can our grief ever change, if the person will never return? Our brain is remarkable, enabling us to imagine an infinite number of future possibilities, if we harness this ability. Chapter 11 closes with what cognitive psychology can contribute to our understanding of grieving as a form of learning. Adopting the mindset that grieving is a form of learning, and that we are all always learning, may make the winding path of grieving more familiar and hopeful.

Think of this book as having three characters. The most important character is your brain, marvelous in its ability and enigmatic in its process. It's the part of you that hears and sees what happens when your loved one dies and that wonders what to do next. Your brain is central to the story, built from centuries of evolution and hundreds of thousands of hours of your own personal experience with love and loss. The second character is bereavement science, a young field full of charismatic scientists and clinicians, as well as the false starts and exciting discoveries of any scientific endeavor. The third and final character is me, a griever and a scientist, because I want you to trust me as your guide. My own experiences of loss are not so unusual, but through my life's work, I hope you may see through a new lens how your brain enables you to carry your loved one with you through the rest of your life.

PART ONE

The Painful Loss of Here, Now, and Close

CHAPTER 1

Walking in the Dark

When I am explaining the neurobiology of grief, I usually start with a metaphor that is based on a familiar experience. However, for the metaphor to make sense, you have to accept a premise. The premise is that someone has stolen your dining room table.

Imagine waking up thirsty in the middle of the night. You get out of bed and head to the kitchen to get a glass of water. Down the hall, you cross the dark dining room toward the kitchen. At the moment that your hip should bump into the hard corner of the dining room table, you feel . . . hmm, what is it you feel? Nothing. You are suddenly aware that you don't feel anything in that spot at the height of your hip. That is what you are aware of—*not* feeling something, something specific. The absence of something is what has drawn your attention. Which is weird—we usually think of *something* as drawing our attention—how can *nothing* draw our attention?

Well, in fact, you are not actually walking in this world. Or, more accurately, you are walking in two worlds most of the time. One world is a virtual reality map made up entirely in your head. Your brain is moving your human form through the virtual map it has created, which is why you can move through your house fairly easily in the dark; you are not using the external world to navigate. You are using your brain map to get around this familiar space, with your human body arriving where your brain has sent it.

You can think of this virtual brain map of the world as the Google map in your head. Have you ever had the experience of following voice directions, without fully paying attention to where you are driving? At some point, the voice tells you to turn onto a street, but you may discover that the street is actually a bike path. GPS and the world do not always match up. Like Google maps, your brain map relies on prior information it knows about the area. To keep you safe, however, the brain has entire areas devoted to error detection—perceiving any situations where the brain map and the real world do not match. It switches to incoming visual information when an error is detected (and, if it is nighttime, we may decide to flip on the lights). We rely on our brain maps because walking your body through your mental map of the world takes a lot less computing power than walking through your familiar house as though it were your first experience—as though you were discovering each time where the doorways and walls and furniture are, and deciding how to navigate each one.

No one expects their dining room table to get stolen. And no one expects their loved one to die. Even when a person has been ill for a very long time, no one knows what it will be like to walk through the world without this other person. My contribution as a scientist has been to study grief from the brain's perspective, from the perspective that the brain is trying to solve a problem when faced with the absence of the most important person in our life. Grief is a heart-wrenchingly painful problem for the brain to solve, and grieving necessitates learning to live in the world with the absence of someone you love deeply, who is ingrained in your understanding of the world. This means that for the brain, your loved one is simultaneously gone and also everlasting, and you are walking through two worlds at the same time. You are navigating your life despite the fact that they have been stolen from you, a premise that makes no sense, and that is both confusing and upsetting.

How Does the Brain Understand Loss?

How, exactly, does the brain walk you through two worlds at the same time? How does the brain make you feel weird when you do *not* bump your hip into the missing dining room table? We know quite a lot about how the brain creates virtual maps. We have even found the location in the hippocampus (the seahorse-shaped structure deep in the brain) where the brain map is stored. To understand what the little computer of gray matter is doing, we often rely on animal studies. Animals' basic neural processes are similar to those of humans, and they also use brain maps to get around.

In rats, we can use a sensor to pick up the electrical signal when a single neuron fires. The rat wears headgear as it runs about, and when the neuron fires, a record is logged of the rat's location when the neuron fired. This gives us information about what landmarks the neuron is reacting to, and where.

In a groundbreaking study by Norwegian neuroscientists Edvard Moser and May-Britt Moser, the rat takes a field trip each day to a box where his neural firing is recorded. There is only one notable thing in the box—a tall, bright blue tower made of LEGO bricks. The rat makes about twenty daily visits to his little box, until the researchers have figured out from his headgear which of his individual neurons fire when he encounters the blue tower. They call these object cells, because they fire when the rat is in the area of the object. Even with the clear evidence that the object cells fire when the rat is near the object, there is still a question of why the neurons fire: Is the neuron firing because it recognizes the sensory aspects of the blue tower (tall, blue, hard), or is it reflecting on another aspect, as in, "Hmm, I've seen this here before"? It would be interesting if the neuron is encoding the history of experience.

So, the researchers took the blue LEGO tower out of the box and let the rat have several more daily visits. Amazingly, there were neural cells that fired specifically when the rat was in the area where the blue tower *used to be*. These neurons were a different group of cells from the object cells, so the researchers called these object-trace cells.[1] The object-trace cells fired for the ghostly trace of where the blue tower should have been, according to the rat's internal virtual map. But what was even more incredible was that these object-trace cells persisted in firing for an average of five days after the blue tower was removed, as the rat gradually learned that the blue tower was not coming back. Virtual reality had to be updated to match the real world, but it takes time.

If someone close to us dies, then, based on what we know about object-trace cells, our neurons still fire every time we expect our loved one to be in the room. And this neural trace persists until we can learn that our loved one is never going to be in our physical world again. We must update our virtual maps, creating a revised cartography of our new lives. Is it any wonder that it takes many weeks and months of grief and new experiences to learn our way around again?

A Question of Maps

Usually, scientists try to come up with the simplest explanation for what they see, and maps are not necessarily the simplest explanation for how we locate things. Another explanation for learning that a blue tower is in a particular location is simple conditioning, a learned association during training. But something more complicated than a learned association occurs, and we know this because of research initiated by neuroscientist John O'Keefe, a mentor to the researchers who found object-trace cells. O'Keefe and Lynn Nadel (now a colleague of mine at the University of Arizona) had a revolutionary idea back in the 1970s.

The scientists designed an experiment to compare the two ideas—learned association versus having a mental map. One hypothesis is that a rat learns where to find food by remembering a series of turns from where he starts to where he finds rewarding vittles. This is cue learning, meaning that the animal is responding to the cues he has seen before, an association. The other hypothesis is that the rat has a map of the world in his brain (more specifically, in his hippocampus), and he finds the tasty vittles by going to where they are on his brain map. This is place learning, as opposed to cue learning.

O'Keefe and Nadel constructed a box with evenly spaced holes where food might appear. When the rat is placed at an entrance to the box, he could easily learn, for example, to turn right and run past two holes and get food at the third hole. But if he is just learning these cues, the same plan won't work when researchers put the rat at a different entrance location to the box. Then if he turns right and runs past two holes, he won't get any delicious vittles at the third. On the other hand, if the rat has an internal map of the whole box, then he does not care which entrance he is placed at initially. He will simply run over to the hole where food appears, knowing the location of the hole in relationship to the entire box.[2]

It turns out that rats have a map for the whole area. The experiment showed that the rats are engaged in place learning rather than cue learning. In fact, individual neurons fire for particular places in the box, a kind of code that represents each location. These individual neurons are called place cells. They help us keep track of where we are in the world, but also where other important things are in the world, like a consistent source of food. Humans, by the same token, have place cells for their refrigerator. It doesn't matter if we come in the front door or the back door to our home, we can head for the refrigerator, using our brain map.

Our loved ones are just as important to us as food and water. If I ask you right now where your boyfriend or girlfriend is, or where you would go to pick up your children, you probably have a pretty good idea of how to locate them. We use brain maps to find our loved ones, to predict where they are, and to search for them when they are gone. A key problem in grief is that there is a mismatch between the virtual map we always use to find our loved ones, and the reality, after they die, that they can no longer be found in the dimensions of space and time. The unlikely situation that they are not on the map at all, the alarm and confusion that this causes, is one reason grief overwhelms us.

Evolution Is a Tinkerer

The first mobile creatures needed to find food, a basic necessity of life. The neural map was probably developed in order to know where to go to fulfill that need. Later, particularly as mammals developed, another need arose: for other members of the species, to care for them, defend them, and mate with them. These are what we call attachment needs. For the moment, let's think of the need for food and the need for loved ones (attachment) as similar problems the mammal has to solve. Now, food and loved ones are obviously different. Food is not always found in the same place, but our loved ones have minds of their own and therefore are even less predictable.

Let's take an example of simple mammals to see how we might still use brain maps as a solution to the problem of locating our loved ones. One of my favorite television shows, *Meerkat Manor*,[3] documents the lives of meerkats in the Kalahari Desert. Meerkats are small rodents that look a bit like prairie dogs. The television show is sort of a cross between *Wild Kingdom* and *The Young and the Restless*. The "Whiskers" meerkat family is headed by a savvy, fierce alpha female named Flower. Each day, Flower and her foraging tribe head out onto the savannah to get the beetles, scorpions, and other tasty items the desert provides for their survival. Some members of the tribe stay home as babysitters and guard the baby meerkats, who are completely helpless. The meerkats search over a huge distance for food and yet return home reliably every night to their tiny babies and their bored babysitters. They know how often to return to an area after they have depleted it of food options. They do all this navigation even though every few days, the meerkats move their entire brood to a different underground burrow. Hundreds of these burrows exist, and the meerkats avoid predators, rivals, fleas, and general housekeeping by moving regularly. The virtual map that these little mammals have in their hippocampus must be vast, and yet they return home over and over again, without any apparent difficulty.

Evolution has endowed social creatures with the computational ability to map out their environment, to know where the good sources of food are, and how soon to return to an area after they have eaten there. But evolution is a tinkerer, and when a new need arises, it uses the machinery available rather than develop a whole new brain system. So, it seems likely that the same mapping encoded in neurons to find food would also be used to map where mammals keep their babies, and how to get back to them at the end of the day. Or how to get back to them in an emergency, like the episode in which Flower races back to the den when she sees a dangerous hawk circling over the burrow where her babies are hidden. As humans, we map where our loved ones are on the virtual map in our head, using three dimensions. The first two dimensions are directly related to the same ones we use to find food—space (where it is) and time (when it is good to forage there). The third dimension I will call closeness. One way to make sure that our loved ones are more predictable is through our bond. The likelihood of finding them increases if they feel motivated to wait for us to come home, or if they have the desire to look for us if we do not. This invisible tether, this bond of closeness, is what British psychiatrist John Bowlby called attachment.[4] Considering closeness as a dimension is a novel idea, and I will tell you more about what I mean by that in chapter 2. For now, let's focus on these three dimensions overall: *here, now,* and *close.*

The Attachment Bond

How do we learn the dimensions of *here, now,* and *close?* When a baby is born, he feels safe and secure when he is in contact with his caregiver. I'll refer to the caregiver as "she" in this section, but there is no reason it couldn't be a father. For contrast, I'll call the newborn "he." During physical attachment to the mother, skin-to-skin contact, the baby is soothed and happy, and he has just enough mental capacity to know the difference between having physical contact and not having physical contact. At this point, the baby does not necessarily know the difference between himself and the person he is literally physically attached to, but there is an innate instinct to cry when that contact is desired. The baby learns that if there is no contact, then crying brings the mother into contact again, with a wonderful soothing result. The baby's brain develops a little more, and now he has a sense of the attachment bond even when there is distance (the space dimension). If the infant can see the mother in the room, or even hear her in the next room, there is a sense that the attachment needs can be met. Here we have the first virtual reality, the mental representation of the mother, based on seeing or hearing cues and not just physical touch. This is the attachment bond bridging across space, like an invisible tether. Mama is just as soothing from across the room, and the baby can get on with whatever the baby would like to do because he feels safe.

Next up, the baby learns about the time dimension. Somewhere in the first year, the baby begins to cry when Mama disappears. Although most people assume this is because of the development of the emotional bond to Mama, there is more to it than that. The baby's brain has to develop in a specific way before that inconsolable wailing happens when Mama leaves. What the baby needs is a working memory. His working memory capacity comes online because of new neural connections between parts of the brain. Now the baby can keep in his mind the memory of what happened thirty to sixty seconds ago (Mama was here), and what is happening now (Mama is not here) and relate the two. Unfortunately, he cannot yet handle the uncertainty of what her absence might mean for him. So, although his brain has matured enough to recognize that the present is an alteration of the past, his only option is to cry out, in the hope that Mama will hear him and come back.

Eventually, with experience, he discovers that although Mama may be gone, she always comes back. The toddler comes to know that he can wait for one *Sesame Street*, or maybe two *Sesame Street*s, and then sure enough, Mama is back and all is right with the world. Now Mama is still present in the virtual reality in the toddler's mind, even when Mama is out of sight and cannot be heard. The attachment needs for love and security are not overwhelming, because the toddler can refer to the soothing knowledge that Mama will return. Thus, the attachment bond tethers them across time.[4]

Space and time have been co-opted from dimensions that the brain had been using to find food. Those mammals that applied these same dimensions to their caregivers survived to pass along their genes. The babies who stayed within sight of their mother survived predators, and toddlers who waited where they were until their mother returned with food got better nutrition and grew strong. Attachment developed because the brain applied a solution from one problem to another problem as the new species of mammals evolved.

When the Dimensions No Longer Apply

Our attachment need—the need for the comfort and safety of our loved ones—requires us to know where they are. As I progressed from being an undergraduate to graduate school, I moved to a new university in a different town. My mother felt very strongly about coming to visit me in my new place. "I need to be able to visualize you where you are now," she said. It helped her to feel closer to me, and I think mapping where I was made her miss me less in my absence.

If we use these three dimensions—*here, now, close*—in the virtual map of our brains to locate and keep track of our loved ones, then death presents a particularly devastating problem. Suddenly you are told (and, on a cognitive level, you believe) that your loved one can no longer be located in space and time. On another level, this does not compute; the brain cannot predict this possibility, because it is outside the brain's experience. The idea that a person simply does not exist anymore does not follow the rules the brain has learned over a lifetime. Furniture does not magically disappear. If the person we love is missing, then our brain assumes they are somewhere else and will be found later. The action required in response to their absence is quite simple: Go look for the person, cry out, text, call, or use any possible means to get their attention. The idea that the person is simply no longer in this dimensional world is not a logical answer to their absence, as far as the brain is concerned.

I mentioned before that we could compare the need for attachment to the need for food. Now, imagine that you wake up one morning and make yourself breakfast, but somehow when you sit down to eat, there's nothing on your plate. No coffee in your cup. You did all the right things, followed the procedures of how to make breakfast, but here's the kicker—in the night, the world has utterly changed, and there is somehow no longer food for you to eat. You order in a restaurant, and the waiter goes away and comes back to serve you, but delivers nothing. This bizarre situation is as foreign as the utter confusion that can occur when we are told that a loved one has died. This confusion is not the same as simple denial, although that may be how others describe it. Instead it is the utter disorientation people experience in acute grief.

Am I Crazy?

The first person I saw in psychotherapy who was dealing with grief was quite certain that she was "going crazy." She was in her early twenties, and her father had died suddenly in a violent accident. She was convinced she had seen him on the street after his accident, wearing the bandanna he always wore, and she could not shake this experience. She truly believed she had seen him, and she also knew this was not possible. Worst of all, she hoped she would see him again, even though she was worried about what he would look like after he had been fatally injured.

Searching for our loved ones after they have died is a very common experience. Holding and smelling their things in order to feel close to them is also very common and does not mean a person is crazy (despite what Hollywood might suggest). What matters is your intention. If you are overwhelmed with missing your deceased husband, and you seek out something to remind you of him, to remind you of the time you spent together, that's one thing. If, years after the death of your daughter, you have kept her bedroom exactly the way it was the day she died, with the same sheets on the bed, untouched since she threw them off while stepping out of bed on that fateful day, and you spend time in the room, trying to re-create your experience before she died, this can be problematic. What is the difference? In the first case, you are in the present and remembering the past, with all the pain and sadness and bittersweetness of having known and loved the person. In the second case, you are trying to live in the past, pretending that time has stopped. And as much as we may hope and struggle and yearn, we will never stop time. We can never go back. We must eventually walk out of that bedroom and be smacked in the face with the present reality.

When the young woman in therapy with me heard that she would not require hospitalization for the vision of her father because she was not "crazy," she became able to talk about her grief. She was able to put into words how much she still needed her dad, because she felt so young and so unsure about what her future held. This yearning, in many ways, is the heart of grief.

Searching in the Night

Religions of the world have long honored this desire to find departed loved ones in the dimensions of space and time. Where did they go? Will I ever see them again? In the wake of a loved one's death, we have an overwhelming urge to reach out to them, and this urge often comes at the very same time when many people turn to religion to understand the meaning of life and their place in the universe. Religions provide answers that soothe and comfort the bereaved. They usually describe a place where the deceased now resides (Heaven, the Buddhist Pure Lands, the Underworld across the River Styx) and a time when we will see them again (Día de los Muertos, the Japanese Obon festival, Judgment Day). In many cultures, people visit the grave or an altar in their home where they go to feel close to the beloved person who has died, to talk to them or ask for advice. The fact that so many different cultures have provided a very concrete answer to *where* and *when* might be an indication that the strong desire to search for and map the whereabouts of our loved ones (the desire to have them *here* and *now*) is biologically based. This biological evidence is embedded somewhere in the brain, if we knew how to look for it.

Of course, the importance of a map of where our loved ones are presents some empirical questions: Do people use the same virtual map when asked where their deceased loved ones are as they do when asked where their living loved ones are? Is this map in the hippocampus? More important, does confidence in the whereabouts of our loved ones, and in our future access to them, provide comfort in the wake of loss? We have no neuroscientific evidence to weigh in on this (yet!). However, a fascinating study looking at the stress response of bereaved individuals and their religious beliefs sheds some interesting light on these questions.

First, keep in mind that when we are upset, our blood pressure goes up, and when we feel comforted, it normalizes. During bereavement, we know that people's average blood pressure rises, as compared to similar people who are not grieving. Sociologist Neal Krause at the University of Michigan has pointed out that when we are repeatedly upset about the loss of a loved one, religious beliefs and rituals may offer a soothing and effective way to help us cope. That soothing response should be visible in blood pressure and in rates of hypertension (high blood pressure that persists over time). Krause devised a clever study in which researchers interviewed older Japanese folks, some of whom had experienced the death of a loved one. Those who were bereaved and believed in a good afterlife did not develop hypertension three years later. They appeared to be protected by this belief. Interestingly, believing in a good afterlife did not predict less hypertension in older Japanese people who were not bereaved. This belief only predicted normal blood pressure for those who were dealing with the stress of bereavement, and who needed the soothing comfort of this knowledge.

It is not a part of a neuroscientist's role to determine whether or not religious beliefs are correct; rather, we are interested in whether or not the way we think about our social ties can affect our physical and mental health. There may be similarities between how the brain deals with one problem (keeping track of our loved ones while they are alive) and another problem (staying connected to our loved ones now that we cannot be with them), according to the brain. Regardless of the veracity of religious teachings, through neuroscience we may be able to understand more about how the brain allows us to experience this awe-inspiring thing called life. Understanding what is soothing, to those who are searching for a deceased loved one, may spark some novel ideas about how to provide comfort to other bereaved people. Perhaps finding ways of providing that soothing comfort would allow their brain, and heart, to rest during this incredibly stressful experience of loss.

Filling In the Gaps

In addition to carrying around wide-ranging virtual maps, another of the marvels of the brain is that it is a remarkably good prediction machine. Much of the cortex is configured to take in information and compare that information to what has happened before, to what it has learned through experience to expect. And because the brain excels at prediction, it often just fills in information that is not actually there—it completes the patterns it expects to see. For example, people can see faces in everything from clouds to toast, by filling in the gaps. We strive to make artificial intelligence that is as good at pattern completion as human beings are. We can even measure this prediction capacity in our neurons. When the brain perceives even a small violation of what it expects, there is a particular firing pattern of the neurons that can be picked up with an electroencephalogram (EEG). An EEG cap of electrodes on the human scalp shows a change in the voltage when the brain detects that the "wrong" thing has happened, milliseconds after it occurs. When your hip doesn't bump into the dining room table when you are walking in the middle of the night, for example, the voltage of your neurons momentarily changes.

Prediction is key to almost all human behavior. We compare the expected sensation of the dining room table at our hip to the lack of feeling we take in through our sensory nerves. However, it is important to note that the brain has already logged what it *thinks* it sensed. Processing sensory information is very quick and is filtered by expectations. When you walked through the space formerly occupied by the dining room table, your brain actually felt the table. *Then* it noticed the difference between the pattern of sensation that it expected and logged, and what actually happened. Imagine the man whose wife has returned home from work at six o'clock every day for years. After her death, when he hears a sound at six o'clock, his brain simply fills in the garage door opening. For that moment, his brain believed his wife was arriving home. And then the truth would bring a fresh wave of grief.

This neural computation of the timing of events is how the brain learns. The Canadian neuroscientist Donald Hebb was famously paraphrased as saying, "Neurons that fire together, wire together."

This means that a sensation (hearing a noise) and the events that follow (my wife walks in the door) trigger the electrical firing of thousands of neurons. When these neurons fire in close proximity, they become more physically connected. The neurons are physically changed. Neurons that are more connected are more likely to fire together the next time. When an experience is repeated over and over, the brain learns to trigger the same neurons each time, so that "sound at 6 p.m." triggers "wife is home."

It requires additional time for you to consult with other parts of your brain that report your wife is no longer alive and could not possibly be opening the garage door. In the meantime, the discrepancy between what you have already logged (your wife is coming in the door) and what you know to be true (your wife has died) leads to the painful wave of grief. Sometimes all this occurs so quickly that it is below the threshold of consciousness, and all we know is that we are suddenly overwhelmed with tears. Therefore, perhaps it is not so surprising that we "see" and "feel" our loved ones after they have died, especially soon after the death. Our brain is filling them in by completing the incoming information from all around us, since they are the next association in a reliable chain of events. Seeing and feeling them is quite common, and it definitely isn't evidence that something is wrong with us.

Additionally, our predictions change slowly, because the brain knows better than to update its whole prediction plan based on a single event. Or even two events, or a dozen events. The brain computes the probabilities that something will happen. You have seen your loved one next to you in bed when you wake up every morning for days and weeks, months and years. This is reliable lived experience. Abstract knowledge, like the knowledge that everyone will die someday, is not treated in the same way as lived experience. Our brain trusts and makes predictions based on our lived experience. When you wake up one morning and your loved one is not in the bed next to you, the idea that she has died is simply *not true* in terms of probability. For our brain, this is not true on day one, or day two, or for many days after her death. We need enough new lived experiences for our brain to develop new predictions, and that takes time.

The Passage of Time

The brain learns whether we intend to learn or not. It does not wait patiently until we say, "Hey, Siri," and then begin encoding whatever happens next. Our brain continuously logs the information received through all of our senses, building up a vast store of probabilities and likelihoods, noting associations and parallels between events. Often this happens without our conscious awareness of those sensations, or of the associations made. This unintentional learning has pros and cons. Because learning is unrelated to our intentions, the brain is learning the real contingencies of the world, even when we are ignoring them or do not consciously notice them. Your brain continues to note the fact that your loved one is no longer present day after day and uses that information to update its predictions about whether they will be there tomorrow. That is why we say that time heals. But actually, it has less to do with time and more to do with experience. If you were in a coma for a month, you would not learn anything about how to function without your husband after you came out of the coma. But if you go about your daily life for a month, even without doing anything someone would recognize as "grieving," you will have learned a great many things. You will learn that he didn't come to breakfast thirty-one times. When you had a funny story to share, you called your best friend and not your husband. When you washed the laundry, you didn't put any socks in his drawer.

So, the brain uses a virtual map to get us around and help us find food, and we have probably evolved to use that map also to help us keep track of loved ones. When we experience a loss through death, our brain initially cannot comprehend that the dimensions we usually use to locate our loved ones simply do not exist anymore. We may even search for them, feeling like we might be a bit crazy for doing so. If we feel that we know where they are, even in an abstract place like Heaven, we may feel comforted that our virtual map just needs to be updated to include a place and time that we have never been. Updating also includes changing our prediction algorithm, learning the painful lessons of not filling in the gaps with the sights, sounds, and sensations of our loved ones.

Keep in mind that the brain cannot learn everything at once. You cannot go from arithmetic to calculus without many, many days of practicing multiplication tables and solving differential equations. In the same way, you cannot force yourself to learn overnight that your loved one is gone. However, you can allow your brain to have experiences, day after day, which will help to update that little gray computer. Taking in everything around us, which updates our virtual map and what our brain thinks will happen next, is a good start for being resilient in the face of great loss.

CHAPTER 2

Searching for Closeness

As children, when we are strongly attached to our caregivers and utterly dependent on them, we learn to understand the role we play in closeness. We realize that some of our behaviors make Daddy mad, and that when he is mad, we dislike feeling disconnected from him. Eventually, we learn to see our actions from Daddy's perspective, and foresee that if we color on the wall, he will not pick us up and hug us when he finds us, crayon in hand. We learn that our behavior is a causal element in the closeness/distance dimension. On the other hand, we also come to discover that our attachment, our closeness, persists despite what we feel in a specific situation. If Daddy is mad at us for crayoning on the wall, he will still save us from the speeding truck if we are playing in the middle of the street. Or, if we have a traffic accident in our parents' car after we first get our driver's license, our parents may surprise us by showing relief and gratitude that we are physically safe, despite the damage we have done. This closeness of attachment often transcends the moment-to-moment emotions they feel toward us, at least in secure relationships. Closeness is partially under our control, and we learn how to maintain and nurture this closeness, but we also trust those who love us to maintain that closeness as well.

Closeness is a third dimension of how we map *where* our loved ones are, in addition to mapping where they are on the dimensions of *here* (space) and *now* (time). I think of this as a third dimension because I believe closeness is understood by the brain in a very similar way to time and space. Psychologists have also called this psychological distance. The easiest way to imagine this concept is in response to the question "Are you and your sister close?" Psychologist Arthur Aron depicted closeness by representing you and the person you love with circles.[1] He called this the Inclusion of Other in Self scale. Considering that he is a scientist, I find that quite a poetic description.

At one end of the scale, the two circles sit next to each other, just barely touching. At the other end of the scale, the two circles are almost completely overlapping, with only small crescents showing at the outer edges to represent the distinct individuals. In the middle of the scale, the circles intersect at their poles. People can reliably indicate how close they are to their loved one by choosing the set of circles that fits their relationship best. In the metric of the overlapping circles, the areas where my best friend and I do not overlap are very small. At the other end of the closeness dimension, psychological distance can be just as powerful. In a room full of family members, you can feel as though you are on an alien planet, with no interest in sharing yourself and no belief that they would understand you anyway.

Being There

Closeness is dimensional in the way that space and time are dimensional. Just like we use time and space to predict when and where we will see our wife or husband next, we can use emotional closeness to predict whether they will "be there" for us. At one end of the closeness dimension, when my partner and I both arrive home in the evening, I may feel confident that I will be able to snuggle into his arms and have him soothe away my terrible day. Alternatively, if our relationship is struggling, the best I may be able to hope for is that we will sit together on the couch watching TV out of habit. If we have recently had an argument, I may give him the cold shoulder, even frowning at him, subliminally warning him to keep his distance.

Since closeness is a metric with which we track "where" we are in relation to our loved ones, the brain struggles with how to understand what has happened when the person dies and this dimension disappears. In the case of space and time, if our loved one is not present, then our brain simply believes they are far away or will be here later. For our brain, it is too unlikely that these dimensions no longer apply, that the person cannot be found in *here* or *now*. When a loved one has passed away, we may feel that we are no longer close, but our brain cannot believe it is because "closeness" no longer applies. Instead, our brain may believe it is because they are upset with us, or that they are being distant. If they are not responding to us, even though we logically know that they cannot, then our brain may believe we are not trying hard enough to reach them, not appealing fervently enough to them to come back to us.

Ghosting

The opposite of closeness is feeling the absence of our partner. Absence sets off emotional alarm bells, revealing the calm and comfort of closeness that we miss. Unexpected absence alarms us even more. A while back, one of my friends developed a long-distance romantic relationship with a guy who lived across the country. Years earlier, they had known each other as friends when they worked in the same place, and they stayed in touch via email after she moved away. Eventually, each of them became single, and their conversations became intimate. They texted daily, intensively. Then one day, without warning, he stopped responding. No email, no texts, no explanation, no idea what had happened. The guy went from intimately close to perplexingly distant in a single night. Ending a relationship by suddenly and inexplicably withdrawing from all communication even warrants its own term in our modern technological world: ghosting.

In addition to feeling profound empathy for the pain my friend was feeling, I was struck by the intense emotional reactions she experienced. She was, of course, deeply hurt and choked up when we talked about it in the days afterward. She also felt raging anger toward him and wrote him several angry emails pointing out that she simply wanted an explanation, and that what he was doing was unbelievably mean. Needless to say, she spent hours considering what might have happened. Had she done something to offend him, even though she couldn't think of what that might be? Did he feel vulnerable after sharing himself emotionally with her, and decide he couldn't face her?

Of course, at some point, we also considered the possibility that a terrible accident had occurred and he had died. Although this turned out not to be the case, I realized something important. When a loved one dies, we may feel many strong emotions in addition to sadness. We feel regret, or guilt, or anger, or what we might call social emotions. On a subconscious emotional level, we may feel that they have "ghosted" us, and we may feel these same intense, motivating emotions of anger or guilt. When our loved one is living, these emotions would motivate us to repair the relationship— to apologize, to fix something that has happened, or to tell them we are upset so that they can make it up to us. But, unlike an argument, when someone dies there is no chance for resolution.

Seeing my friend going through this painful breakup brought home a vital point. If your brain cannot comprehend that something as abstract as death has happened, it cannot understand where the deceased is in space and time, or why they are not *here*, *now*, and *close*. From your brain's perspective, ghosting is exactly what happens when a loved one dies. As far as the brain is concerned, they have not died. The loved one has, with no explanation, stopped returning our calls—stopped communicating with us altogether. How could someone who loves us do that? They have become distant, or unbelievably mean, and that is infuriating. Your brain doesn't understand why; it doesn't understand that dimensions can simply disappear. If they don't feel close, then they just feel distant, and you want to fix it rather than believe they are permanently gone. This (mis)belief leads to an intense upwelling of emotions.

Anger

Sadness is probably the easiest feeling for us to comprehend during grieving. Something is taken away from us, and it is not hard to imagine that this would lead to sadness. If you take a toy from a toddler, or a toddler's mother leaves, it makes perfect sense that his little face breaks up and he sobs as though his heart would break. Sadness makes sense. But I have always found the strength of the anger that we experience during grief to be notable and somewhat perplexing. Why are we so angry? Who are we angry at? Sometimes our anger is directed at the person who died. But we can be angry toward a range of people, including doctors and even God. This anger is motivated by something different than anger we feel toward the person who died. If you take a toy away from a toddler, he may scream at you in anger. And sure enough, sometimes you give him the toy back, because you see how much it has upset him. But no one can return the person who died.

Not being able to sense our loved one who has died and feeling on some level that they are ignoring us, calls everything we believe into question. Like my friend and I did during phone calls after she was ghosted, we run endless possible scenarios after a death. How could this have happened? Could we have stopped it? In fact, people who are grieving very commonly describe endless rumination. This "would've/could've/should've" loop can feel exhausting.

During grief, we are not sad or angry simply as a reaction to what happened, the way we would be if a possession were taken away from us. In some cases, we are sad or angry at ourselves because we have "failed" to keep our loved ones close on the closeness dimension. This failure on our part, or on their part, is upsetting in all sorts of ways. It does not have to make logical sense that our brain believes the person has ghosted us. We can know that it is ridiculous to be angry with the person for dying, or futile to be angry at ourselves for not keeping them close, and at the same time be furious anyway. Just as the brain may sometimes believe that our deceased loved one is out there, and we may feel motivated to search for them, the brain can also believe that by repairing our relationship with them, we can somehow bring them back.

Evidence of the Close Dimension in the Brain

Psychologists and neuroscientists have been studying how different metrics of *here*, *now*, and *close* could be encoded in the brain. One theory proposed in 2010 by psychologists Yaacov Trope and Nira Liberman at Tel Aviv University is called the construal level theory.

The theory says that when people are not currently present in one's immediate reality, they could be gone for a few different reasons. These reasons include distance, time, and social closeness.[2] We can form abstract ideas, or construals, of where they are or might be. So even though we are not directly experiencing someone through our senses, we can use predictions, memories, and speculations to imagine the person. These mental representations transcend the immediate situation.

The construal level theory also suggests that the brain uses different dimensions to produce reasons for the absence of a person (distance, time, and closeness), just as I have been applying the concept of dimensions to tracking our living loved ones. Because our mental representation of our parent or spouse includes the dimension that they are psychologically close, we can apply this knowledge to making predictions. We can confidently predict that if they are not where we expect them to be, they will be motivated to give us a call, or to show up at home. On the other hand, we do not predict this behavior for people with whom we are not close. We do not expect the head of the company we work for to call us if he does not show up to work. If we have not been to our regular coffee shop for a while, we do not expect a barista to get in touch.

The construal level theory suggests that the brain similarly encodes these dimensions of *here*, *now*, and *close*, and that we even use language to describe these dimensions in interchangeable ways. For example, if I describe something as being "way out," I could be equally understood to mean something far off in time (that appointment is still way out), far off in space (the ball is way out in the field), or someone who is psychologically distant or not relating well to other people in the group (that guy we met today seemed way out).

A couple of neuroimaging studies from the 2010s support the idea that the brain may have a region that computes these different types of dimensions in a similar manner. To demonstrate this, participants looked at photos while they were in the MRI scanner.[3] One set of photos showed a bowling ball at different distances down an alley. Another set of photos showed words used to describe time, such as "in a few seconds" and "years from now." A final set of photos showed close friends and mere acquaintances of the person being scanned. After people looked at photos from each of the three sets, they made judgments about how far away things were. Remarkably, the same part of the brain was used to compute the difference between the pairs of photos that were "near" and "far." For those of you who are brain region junkies, this region is the right inferior parietal lobule (IPL). That means that neurons encode different distances, and the brain uses that common code for proximity to the self regardless of whether it is considering time, space, or psychological closeness. You might think it would make more sense for the brain to consider time in one brain region, space in another region, and psychological closeness in a third. But apparently it is more efficient for the brain to represent the distance aspects in the same computational region, since they carry a common metric.

Another fascinating and clever neuroimaging study by neuroscientists Rita Tavares and Daniela Schiller looked at how psychological closeness is encoded by the brain. Tavares scanned people's brains while they played a choose-your-own-adventure game.[4]

You may remember reading choose-your-own-adventure books when you were a kid. You got to choose what you, as the main character, would do next in the story (within a limited set of given options), and then you would turn to the page of the choice you made for the story to continue. In the case of Tavares's neuroimaging study, each person being scanned in the study played the role of the main character. In one scenario, a new friend, Olivia, suggests that you do the driving on this adventure. You could choose to take the driver's seat while she gives you directions. Or you could decide you don't trust Olivia enough to give you directions, and since you don't know your way around, you could suggest that she drives. In another example, Olivia offers you a hug, and you may choose to give her a pat on the back in return, or to hug her for a long moment, based on the closeness you have developed during the story.

The psychological closeness dimension was measured from the study participant (the main character) to the other characters in the game, quantifying how close the person being scanned felt to the people in the story. The level of closeness evolved during the scan, as the story unfolded based on the decisions made by the person being scanned. The researchers then used geometry to calculate the change in how close the participant felt to each of the characters over the course of the game. As the participant developed a closer relationship to another character in the game, the researchers could compute the shrinking distance. Amazingly, the results of the study confirmed the scientists' predictions. A part of the brain was literally tracking which characters became part of the participant's "inner circle," or surpassed their own status and became more distant as they went "up the corporate ladder," by the end of the game. The region of the brain that measures the amount of *closeness* between people is the posterior cingulate cortex (PCC), a region that I will tell you more about in chapter 4. In other words, the psychological distance between the participant and the characters was encoded as a neural firing pattern in the PCC. In addition, the hippocampus tracked "where" the character ended up in this social space, utilizing the unique capacity of the hippocampus for social navigation, similar to the way it maps spatial navigation. Even as a neuroscientist, I am astounded at the ingenuity of the brain in developing a neural map that tracks how close we feel to people, even in an abstract space.

This study provides evidence that the ephemeral sense of closeness with our loved ones exists in the physical, tangible hardware of our brain. A change in our feeling of closeness with others arises in the posterior cingulate cortex, and is delivered to our conscious awareness. Like an intelligence analyst, the PCC absorbs hundreds of small bits of information from the brain's sensory agents in the world. Like a team of police detectives with red strings between suspects on an investigation board, the PCC constantly updates for the connections between ourselves and others, shortening the strings as we feel closer to another person, lengthening the connections when detecting more distance. After the death of a loved one, the incoming messages seem scrambled for a while. At times, closeness with our deceased loved one feels incredibly visceral, as though they are present in the room, *here* and *now*. At other times, the string seems to have fallen off the board—not shorter or longer than it was before, but simply stolen from us entirely.

Closeness and Continuing Bonds

Closeness in your relationship with your loved one is transformed after they die. That transformation works differently for different individuals, since each of our relationships are unique. Psychiatrist Kathy Shear at Columbia says that "grief is the form love takes when someone we love dies."[5] Many cultures emphasize relinquishing the bond with the loved one as a part of facing the reality that they are gone. Some cultures emphasize that the bereaved should continue the relationship and communicate with the loved one, or even have rituals through which they are transformed into a continuing presence as an ancestor. Psychological science calls these continuing bonds. These bonds are unique to each relationship, and the people we have interviewed for research have graciously shared some of their intimate moments. One example came from a young woman whose husband had died. The couple had shared a love of music, and she continued to feel connected to him through the songs she heard. She recalled driving home one afternoon, and every song that came on the radio seemed to be related to him in some way. Her vision of him DJ-ing her ride home made her laugh, and the continued connection consoled her.

At one time, Western clinicians believed that continuing bonds were a sign of unresolved grief and that severing this connection to an inner dialogue with the deceased allowed us to create stronger bonds with our living loved ones. More recent research has shown that although wide variation exists in these inner relationships, many people adjust well by maintaining a connection to the deceased. A widowed woman told me that when she spoke to her teenage son, she felt her deceased husband was helping her to find the right words to say. Another woman told me about writing letters to her deceased husband, asking all sorts of questions about what she should be doing and how. Continuing bonds occur not just through conversations; they may include carrying on the wishes or values of the deceased. No research has investigated yet whether the closeness of these continuing bonds can be mapped in the brain. Someday we may have an answer to how this type of closeness works at the neural level.

The Ties That Bind

Attachment bonds, and the resultant continuing bonds, are the invisible tethers that motivate us to seek out our loved ones, and to derive comfort from their presence. We develop these bonds with romantic partners when we fall in love. The neurochemistry in our brain and our body stimulates, and is stimulated by, falling in love. Another way to think about falling in love, or entering a long-term relationship with another person, is the process of overlapping our identities. Including the other in the self, we become overlapping circles.

You might even think of this as the merging of resources, so that we come to feel that what is mine is yours, and what is yours is mine. The enduring nature of bonds, such as pair bonds, separates an attachment relationship from a transactional relationship. In a transactional relationship, such as with a colleague or an acquaintance, we track whether we are putting more effort, time, money, or resources into the relationship than they are, and how much we are getting out of it. With attachment, both people have access to help at the times when it is needed most. Examples include support and caregiving when one of you is ill, giving the other person the benefit of the doubt, or defending the other's reputation. In a healthy and mutual relationship, we engage in these behaviors not because something equal will be gained in return, but because these are expressions of love and caring. In fact, research shows that providing selfless support has health benefits for the provider as well as for the beneficiary.

As a concrete example of merging resources, when two people have lived together for a long time, there is no longer the question of who owns the sofa. But I am not referring just to things. We also feel other overlaps. For example, we do not necessarily remember who came up with the idea for a wonderful trip we took together, an experience that we both enjoyed. We may confuse which one of us said something particularly witty in a conversation, when we re-tell the story later. The overlap of our resources is an overlap in our identities, as "we" becomes more important than "you" and "me." Falling in love is accompanied by the rapid expansion of these resources, although we might not consciously describe it that way, and the expansion is a pleasurable and exciting feeling. By the same token, there is a correspondingly intense negative contraction after the loss of a person. You may wonder who you are now, or what your purpose is, without the other person. If your child has died, are you no longer a mother? Or it may feel like you cannot go on without your partner. You may feel at a loss for what to do in situations where you previously decided things together. Unable to share your day's events when you get home in the evening, you might feel almost as though they never happened.

Grief emerges as distress, caused by the absence of a specific person who filled one's attachment needs and therefore was part of one's identity and way of functioning in the world. We can look at other situations that also produce grief and see that they share some aspects of this definition. The loss we experience through divorce (or a breakup) is clearly very similar. The loss of a job, through retirement or being laid off, is a loss of the identity that has helped you function in the world. The loss of one's health, the loss of a limb or one's sight—all these are losses of function, but are also experienced as losses of part of who you are. Although I believe that grief in the brain's neurochemistry originally evolved specifically to cope with the death of a loved one, these other similar situations piggyback on that evolved capacity, and we recognize the internal experience as grief.

Grief Over Famous People

If grief distresses us because of the loss of closeness, then why do we feel such an outpouring of grief over the death of a famous person whom we never knew personally? Michael Jackson died at the UCLA Ronald Reagan hospital, just a block from my office at the time. You may remember that afterward the sidewalk by the hospital was littered with flowers and stuffed toys and cards. More recently, the untimely death of actor Chadwick Boseman prompted an unprecedented outpouring of grief online. Given what I have said about attachment (and bonding) being critical to grief, it seems counterintuitive that people would experience such intense mourning after the death of a person they never knew, never encountered in real life.

This type of grief is *parasocial grief*; it is very real, and it goes beyond anecdotal evidence of people who have felt bereft over a celebrity's death. People are represented in the virtual reality of our brains, and celebrities can have very fleshed out lives in our minds. We have a surprising amount of access to what famous personalities portray as their lifestyles and beliefs, their friendships and romances, their likes and dislikes. This kind of information is not necessarily sufficient for forming an attachment bond; however, if we think about what the prerequisites for attachment are, our relationships with famous musicians and celebrities may still meet the criteria to some degree. First, the person must meet our attachment needs. This means that the person is available when we need someone to turn to in our darkest hour. Who has not binge-watched a show with a favorite actor (for me, Gillian Anderson), as a break from the painful reality we are dealing with? I carried around the cassette tape of "Little Earthquakes" to play in my Walkman whenever I felt lonely or sad or overwhelmed for years. The time spent in communion with that famous person—in an emotional state, and possibly enhanced by dancing and screaming in the midst of a like-minded group, or even by alcohol and drugs—can mimic the time spent in attachment bonding.

Attachment requires another aspect, however, other than believing that the person will be there for us. The person also has to seem special, different from other people, our special one. After Michael Jackson died, a friend told me that growing up as a young Black man in the '80s, you were either a Michael Jackson guy or a Prince guy. Endless debates in the halls of the high school ensued about which one was better, but at the end of the day, you belonged to one camp or the other. We choose the celebrities we love, who we identify with, who we believe in as the most talented, sexiest, or best. We often feel close to musicians—we feel we can trust them, because they say what no one else says in their lyrics. They are "yours," in a way. And it also feels a little as though they know us, too, because they say the things that we feel deep down and don't admit to anyone else. How could they write those lyrics if they did not deeply understand you, if they were not speaking to you directly? The loss of that celebrity is not only the loss of a person who helped to define us, but also grief over a time in our lives that we can never return to. That grief is real because we feel the loss of a piece of ourselves.

Losing a Part of You

One of the questions I ask, when I am sitting across the table interviewing a bereaved person for a research study, comes from a psychological scale measuring the severity of people's grief. I will never forget one woman's reaction to a particular question. I asked, "Do you ever feel that a part of yourself died along with your husband?" Her eyes got big and she stared at me, with a look that said, *How could you possibly know?* "That's exactly how I feel," she replied.

If psychological closeness can make us feel so close as to be overlapping with another person, the brain must process this, and compute the overlap of another with one's own self. Think of driving down a road with multiple lanes of traffic. You drive in the middle of the lane—except that description is not quite accurate. After all, you do not put your body in the middle of the lane, because then the car would be more in the lane to your right. Experienced drivers learn pretty quickly to extend their "body" to encompass the whole car. We feel as though we are driving in the middle of the lane, but really, we are centering the car in the lane, and our body is off to the left, even though we do not feel this consciously. In our mind, the car and our body are overlapping. When we have this experience, the brain is computing this overlap.

Grieving people often describe having lost a part of themselves, as if they have a phantom limb. Phantom limb sensations happen in many people who have a limb amputated. Even though their arm is missing, for example, they continue to have the sensation that it itches. Once believed to be an entirely psychological phenomenon, studies have proven that the sensations are actually nerve activity. Researchers believe that the part of the brain that contains a map of our body no longer corresponds to the peripheral nerve sensations.[6] Thus, despite a lack of sensory nerves actually firing in the phantom limb, the brain map has not yet rewired itself, has not updated to dismiss this part of the body, so the sensations persist and are often painful.

We might think it is simply a metaphor to say that we have lost a part of ourselves when a loved one dies, but as we have seen, representations of our loved ones are coded in our neurons. Representations of our own bodies are coded in our neurons as well, as demonstrated by the phantom limbs. These representations of the self and the other, this closeness, is mapped as a dimension in the brain. Consequently, the process of grieving is not just about psychological or metaphorical change. Grieving requires neural re-wiring as well.

Mirror Neurons

Evidence for closeness includes an overlapping neural coding of self and other. This evidence has been concretely demonstrated through another set of scientific studies. Aptly named mirror neurons are designed to fire both for our own actions and for someone else's actions. In the 1990s, they were discovered in the premotor region of the brain, although they have been found in some other regions as well. This overlap in neural firing patterns for self and for another can be seen during mimicry.[7] If you show a monkey that you are doing something with your hand—grasping a banana, for example—some of his same neurons will fire when he watches you grasp the banana as when he grasps the banana himself. Put differently, the neurons that fire when we execute an action of our own vicariously fire while watching the same action by another.

Despite the widespread interest in mirror neurons, human neuroimaging does not have sufficiently high definition to detect individual mirror neurons in humans. In human neuroimaging, we look at brain regions, or populations of many neurons, whereas in monkeys, we are able to detect the firing of individual neurons through invasive recording methods. That said, there has been one report of mirror neuron activity from the electrical recording of a neurosurgical patient. Even with such minimal evidence in humans, we have no reason to believe that a neural system would work completely differently in such closely related primates as macaque monkeys and humans.

No matter how close we are to another person, we are still able to distinguish between the self and the other. In a study examining primates, two monkeys each held their own banana. Imagine a Venn diagram representing neurons in Monkey 1's brain. The circle on the left represents the neurons that fire when Monkey 1 thinks about holding his own banana, and the circle on the right represents the neurons that fire when Monkey 1 thinks about Monkey 2 holding her banana. These circles overlap a little bit, meaning some of the same neurons fire both when Monkey 1 thinks about himself holding a banana and when he thinks about Monkey 2 doing the same thing. But there are also portions that do not overlap. This means that Monkey 1 is able to distinguish his own action from her action, even while overlapping neurons indicate evidence of overlapping identity and shared experience, the particular type of closeness we also see in humans.

Empathic Concern

Neural machinery enables us to feel close to another person, and that machinery includes mirroring others' actions by feeling those actions as though we were performing them ourselves. I have been using these neuroscientific findings to explain how we might feel overlapping with a loved one, and what happens when that person dies. But we can also extend this to the idea of being "grief adjacent," or for how we feel when we are around someone who is grieving. When a friend is grieving, when they are learning to adapt to feeling that a part of them is missing, it affects those who care for them, often deeply.

You may be surprised to hear how contagious sadness can be. We can feel the emotions that someone else is feeling, by simulating that same feeling in ourselves. Science has shown this by investigating the eyes, as they are windows to emotional states, if not to souls. In a study done by British psychiatrists Hugo Critchley and Neil Harrison,[8] student volunteers were shown pictures of faces with happy, sad, or angry expressions. Although the students did not know this, the pupil size of the eyes in these pictures was digitally altered to vary from small to large (within realistic biological limits). Students rated the sad expressions as more intensely sad when the pupils of the pictured faces were very small. More important for thinking about contagion, the different-size pupils had a large impact on some students' ratings of sadness intensity. Those who were very sensitive to the differences between eyes were also higher on measures of empathy. And the more pupil constriction there was in the pictures of the sad faces, the more the students' own pupils constricted when measured with a pupillometer. This type of emotional contagion, such that the pupils of an observed person can affect the emotional experience and physiology of the observer, can happen even when the observer is not consciously aware of it. The students did not know that their own pupil size was changing in response to the photos. We appear to be hardwired to be influenced by the people around us, to be sensitive to cues of what they are feeling—in other words, we are hardwired with the neural building blocks of closeness.

Emotional contagion can be a bad thing. Just like the monkey who would not know who was grasping the banana if they only had mirror neurons, feeling what everyone close to us is feeling could be overwhelming and could cause you to withdraw from them if they are sad or angry. However, scientists now make a distinction between empathy and compassion. In addition to being sensitive to what others are feeling, compassion is defined as also having the motivation to care for their well-being. As neuroscientist Jean Decety from the University of Chicago explains, there are actually three aspects to empathy. These are *cognitive perspective taking*, *emotional empathy*, and *compassion*.

The cognitive aspect of empathy is the ability to see or imagine another person's perspective, unrelated to their feelings. If you are sitting face-to-face with someone, you know that they cannot view the scene that you see behind them. But, because you can take their perspective, you understand that if someone walks into the room behind them, the person across from you is unaware of this. You would have to tell them this person has arrived. That ability to take someone else's perspective is an example of the cognitive aspect of empathy. Emotional empathy, on the other hand, is being able to feel the way another person is feeling. For example, if you and your friend are both in line for the same promotion and you get it, you may put yourself in your friend's shoes and feel their disappointment, despite feeling happy for yourself. And compassion, or caring, goes beyond empathy. It is the motivation to help or comfort the person when you can take their perspective and know how they are feeling.

When a bereaved person has lost the dimensions of *here, now,* and *close*, their emotions may be intense, or they may feel numb.

Compassion from a friend who is grief adjacent will not fill the hole where their deceased loved one has been torn out of their overlapping sense of "we." But it will place supports around the hole, while your friend begins to restore her life. It will help her at least to get through the confusion about what has happened as her life is turned upside-down, which is the topic we will turn to next.

CHAPTER 3

Believing in Magical Thoughts

A few years ago, an older colleague of mine passed away. I spent some time with his widow in the months afterward. As a prominent sleep researcher, her husband had traveled quite often to attend academic conferences. Over dinner one night, she shook her head as she told me it just did not feel like he was gone. It felt as though he was just away on another trip and would walk through their door again at any minute. We hear this kind of statement quite often from those who are grieving. People who say this are not delusional; they simultaneously are able to explain that they know the truth. They are not too emotionally frightened to accept the reality of the loss, nor are they in denial. Another famous example of this belief comes from Joan Didion's book *The Year of Magical Thinking.* Didion explains that she was unable to give away her deceased husband's shoes, because "he might need them again." Why would we *believe* that our loved ones will return, if we *know* that's not true? We can find answers to this paradox in the neural systems of our brain, systems that produce different aspects of knowledge and deliver them to our consciousness.

If a person we love is missing, then our brain assumes they are far away and will be found later. The idea that the person is simply no longer in this dimensional world, that there are no *here, now,* and *close* dimensions, is not logical. In chapter 5, I will tell you more about the neurobiology of why we *want* to find them. For this chapter, however, the question to consider is, Why do we *believe* we will find them?

Evolutionary Contributions

Psychologist John Archer, in his book *The Nature of Grief,* pointed out that evolution has given us a powerful motivation to believe that our loved ones will return, even when the evidence says otherwise. In our early days as a species, those who persisted in the belief that their mate would return with food stayed with their young. The young of those parents who waited with them had a better chance of surviving. We observe this phenomenon in the animal kingdom. In *March of the Penguins,* we see a father Emperor penguin incubating his egg at the inhospitable South Pole, while the mother goes foraging in the icy sea. His motivation to stay with the egg is remarkable: the male penguin fasts for about four months waiting for his partner to return. As an aside, I should mention that same-sex pair-bonded penguins have turned out to be equally good parents. Pair-bonded male penguins Roy and Silo at the Central Park Zoo incubated and raised a sweet baby penguin named Tango.[1]

Regardless of who the parents are, the key here is that one parent must persist in the belief, during a very long absence in the Antarctic, that their partner will return with food. If one parent decides that their partner will not return, and goes to the sea to fish, then the egg fails to hatch, or the chick dies. Those penguins who persist in the belief their partner will return, and wait for them, are far more successful. In the film we see that among thousands of penguins, the returning mother finds her partner by recognizing his very specific call. It is a remarkable phenomenon, with these animals overcoming seemingly endless odds.

What is it that allows the brooding parent to stay on the egg, fasting for months? What is the mechanism of this attachment, or what creates the invisible tether between the pair? The bonding between the parents is compelling. Earlier in the season, the lovebirds spend time with their necks entwined, vocalizing sweet nothings to each other. Simultaneously, their brains are undergoing a physiological transformation. The neurons are stamping the memory of this particular penguin, tagging the neurons with markers that mean it is unlikely that it will ever forget the way this specific penguin looks, smells, and sounds. In the brain, the partner goes from a recognizable penguin to *the* penguin of great importance. Throughout the time that the penguins spend apart, brooding on the egg, the memory of the other is not just a memory. It is a memory attached to a specific belief or motivation—"Wait for this one to return. This one is special. This one belongs to you." In humans as well, it is *because your loved one existed* that certain neurons fire together and certain proteins are folded in your brain in particular ways. It is because your loved one lived, and because you loved each other, that means when the person is no longer in the outer world, they still physically exist—in the wiring of the neurons of your brain.

Primate Grief

Although *March of the Penguins* is a vivid and useful example of what it looks like when creatures persist in the belief that one's loved ones will return, a Disney movie is not the basis of scientific evidence. After all, we aren't descended from penguins. Another way to look at evolutionary evidence is to look at the behavior of those who share a common ancestor. Chimpanzees are humans' closest living relatives, as both species descended from a common ape ancestor.

Several chimpanzee communities around the world have become the source of scientific observation, including the famous chimps of Gombe that were documented by Jane Goodall, and the chimpanzees of Bossou studied by researchers from the Kyoto University Primate Research Institute. In reaction to the death of an infant, these highly evolved mother chimps carry their baby for days after it has died. Chimpanzee mothers (and, in other cases, apes and monkeys) continue to carry and groom their infants after their death, from a few days up to a month or even two. This has been documented dozens of times, with extensive observations of who, when, where, and how. A chimp mother named Masya carried her infant for three days, often peering intently into the face of her baby.[2] She continued to groom the little one, carefully carrying the lifeless infant even when that made it difficult to eat and move around. Carrying is actually unusual behavior for mothers, because baby chimps usually cling to them, which frees the mothers' hands for other activities. During this time, Masya stopped interacting with her troop and didn't groom herself at all. She never tried to nurse the baby, suggesting that she knew that the infant was no longer alive. In a compassionate response from the community, other chimps from the troop began to groom Masya, as she focused intently on her infant. Gradually her behavior shifted from constant contact and protection to finally being able to leave the infant's body permanently. In a different situation, when an infant chimp died of a potentially communicable disease, the researchers removed the corpse after four days. Afterward the chimpanzee mother searched for the baby, vocalizing the whole time. This behavior is not seen when the mother is allowed to relinquish the infant in her own time.

Spending these days with the corpse of the infant, the mother chimp experiences the death of the infant in no uncertain terms. In this way, the belief that attachment creates, the magical thinking that this special one will always be there, is disproven by the mother's own experience. It is likely that human cultural events like funerals, wakes, and memorials serve a similar purpose. Preparing for a memorial includes calling family and friends, and telling them of the death, and hearing their condolences. I remember awakening the morning after my father died, and our dining room table was covered with a dozen flower centerpieces my sister had created for the tables for his memorial. I could sense that the act of creating them, the time it took to choose vases and add ribbons, was a part of her processing the fact of loss. When family and friends travel many miles, put on special clothing, and join together to give hugs and smiles and love—these all mark the moment as different, and that moment stamps in our memory the fact of the death. At many funerals we see the corpse of our loved one in a coffin, or see an urn of ashes, the physical proof that their bodies are no longer the vessels for the souls that we love. A community recognizes, and shows explicitly in their behavior, that this person is not going to return. It reinforces what the bereaved survivor can only half believe at the time. Afterward, when we have memories of the funeral, these memories may help us a little in untangling our own magical thinking; as hard as it is to believe, memorials are proof that others share our new understanding that our loved one is gone.

Memories

If we take seriously what bereaved people tell us, then it seems the brain can persist in two mutually exclusive beliefs. On the one hand, we have the clear knowledge that a loved one has died, and on the other, the simultaneous magical belief that they will return. When a loved one has died, we have a memory of learning that they died. This memory might be of the phone call informing you that your brother died, etched in your mind with lots of detail—where you were in the dining room, what you were cooking, how hot it was in the room, the smell of onions. These are what we call episodic memories; they are detailed memories of a specific event.

Perhaps your memory of the death occurred because you were there when it happened. When my father passed away in the summer of 2015, my sister and a dear family friend and I had been taking turns sleeping in the room with him in the hospital he had chosen for hospice care. On that particular night, I had said good night to him, although he was not responding to us anymore. I got a few hours of sleep on the little couch in the room. In the middle of the night, I woke up filled with a sense of awe, a frequent experience in those last few days (along with feelings of utter exhaustion and lack of confidence that I could go on any longer). I checked on my dad, and then I decided to go for a walk outside, moved by a similar sense of awe I feel when looking at the marvelous stars in the rural Montana night sky. If you've ever been far, far away from city lights, you know that there are so many stars, the night sky looks like it is scattered with glittering sand. I walked the circular path around the hospital, designed to give staff and visitors a place to stretch their legs. I went back to the room, and Dad was still breathing very, very slowly. It was truly amazing, I thought, that his life could be sustained with so little breath. I went back to sleep. In the wee hours of the morning, a nurse leaned over me, hand on my shoulder. "I think he's gone now," she said. I went to my dad's bedside. He was so peaceful, so little, looking both like an infant and an old man at the same time. He looked exactly the same as a few hours before, except that his breaths had gone from very, very slow to none at all.

My experience of my father's death was extremely peaceful and filled with awe, and I was comforted by loved ones and caring professionals around me. I was able to really focus on what was happening at the time, and as I look back on it I usually feel quite peaceful, even if very sad. I count myself extraordinarily lucky, because I got to experience what could only be called a good death. It was aided by the fact that he was in a hospice program, designed by the people who know the most about creating conditions most likely to lead to a good death. Many deaths are not at all like this. People experience fear, terror, pain, helplessness, or extreme anger at the moment of their loved one's passing, especially if it occurs in violent or terrifying circumstances, in accidents or emergency rooms. During the COVID-19 pandemic, many people were unable to be with their loved ones when they were admitted to the hospital and were not at their bedside when they died. Without the opportunity for saying good-bye, for expressing love, gratitude, or forgiveness, and without the memory of seeing our loved one's physical decline and death, ambiguity may surround the "realness" of the death. Research shows that ambiguous loss, such as when family members are disappeared by a political regime or missing and presumed dead from an airplane crash or in wartime conflict, complicates the grieving process. One reason may be that part of our brain is wired to believe that our loved one is never really gone, and without the overwhelming evidence from our memories of their decline or death, rewiring our understanding may take longer or cause greater distress.

Habits

Memory is extremely complex. Fortunately, it is also an area that many neuroscientists and cognitive psychologists have been studying for a long time, and so we know a fair amount about how it works in the brain. The brain doesn't function like a camcorder, recording every moment of every day and then storing it forever. It is easy to imagine that memories are like a video stored in a file folder that the brain opens and plays when we remember something. Actually, memories function a little more like cooking a meal. The ingredients of our memories are stored across many areas of the brain. When we remember an event, these ingredients are brought together, dumping into the mix sights and sounds and smells, a feeling the event created for us, associations with particular people at the event, the perspective from which we viewed various scenes. Together, memories appear to us as a synthetic experience of an event in the past, as a cake appears to be a single entity and not a combination of flour, sugar, and eggs. However, different cakes have different flavors, like chocolate and vanilla, although they are still identifiable as cakes. Similarly, whether we are in a good mood or a bad mood when we recall the memory affects the ingredients we include in this version of the memory, perhaps making our recollection brighter with color or more bittersweet. Sometimes when I remember my father's death, my memory is not dominated by the amazement I felt, and instead my exhaustion is the primary memory. And even though I am not entirely certain whether the nurse put her hand on my shoulder or simply woke me by speaking, the episodic memory is still recognizable to me as it unfolds in my mind.

Memories enable us to learn from situations that we have experienced, and a significant event like the death of a loved one is likely to be prioritized in the brain's database. You could think of episodic memory as one type of knowledge, knowledge of specific events or moments, accessed by the brain because of its importance in your life.

C.S. Lewis, the author of *The Chronicles of Narnia*, also wrote a poignantly insightful book entitled *A Grief Observed*, after the death of his wife. In it, he writes:

> I think I am beginning to understand why grief feels like suspense. It comes from the frustration of so many impulses that had become habitual. Thought after thought, feeling after feeling, action after action, had [my wife] for their object. Now their target is gone. I keep on through habit fitting an arrow to the string, then I remember and have to lay the bow down. So many roads lead thought to H. . . . So many roads once; now so many cul de sacs.

It is common during grieving to repeatedly recall a very important episodic memory, like the sound of a voice on the phone telling you that your brother died, or the sight of your father no longer breathing in his hospital bed. As part of your brain plays out the memory, another part of your brain is summing up the new experiences caused by his absence and developing new predictions, new habits, new routines. This knowledge contrasts with the magical belief that our loved one is somewhere, just not *here, now*, and *close* at the moment.

Two Beliefs That Are Mutually Exclusive

It may be the cruelest aspect of our human nature that we can experience these incompatible mutual beliefs—both that our loved one is gone and that they can be found again. During all this, our brain holds a persistent representation of the other person, or an avatar of our beloved, in our brain's virtual world. The encoding of this representation emerges while a parent nurses a child, or during a couple's intimate moments. Inherent in this representation of our one-and-only, as a consequence of attachment, is that we believe so thoroughly in that person's existence that we create a never-ending relationship to them, the persistent belief in *here, now,* and *close.* The neural connections that serve as the algorithm for the mental representation of our loved one are permanently encoded. Our plans, our expectations, our beliefs about the world are influenced by this implicit knowledge, our belief that our loved one will return or can be found. Implicit knowledge might be blamed for our magical thoughts.

Implicit knowledge, operating below the level of consciousness, influences our beliefs or our actions. How do scientists know that implicit knowledge exists, if it operates below the level of consciousness? If the person cannot report his knowledge, then we can only see the effect of this knowledge on people's actions. But a convincing piece of evidence that the neural machinery creates implicit knowledge comes from neuroscientific studies of people who have suffered damage to specific parts of their brain. A famous patient, Boswell,[3] was unable to form any new memories because of an accident that caused damage to the temporal lobe of the brain that contains the hippocampus and amygdala. This type of memory deficit, the loss of the ability to create new memories, is called anterograde amnesia. He could not recognize anyone whom he had met in the fifteen years since his accident, even those with whom he had daily contact.

However, Boswell still had implicit knowledge about people, which was revealed through closely studying his behavior. The researchers realized that Boswell gravitated toward a particular caregiver, showing a preference for him over other staff, despite not being able to recognize him or being able to tell the researchers this caregiver's name. Although he had no episodic memory of when, where, and under what circumstances he had met this caregiver, he seemed to be drawing on other knowledge to form a preference for him. The researchers also noted that this particular caregiver behaved very kindly toward Boswell and frequently gave him treats.

To create controlled conditions to demonstrate that Boswell had implicit knowledge despite his brain damage, the researchers, Daniel Tranel and Antonio Damasio, asked him to do a special kind of learning task. They introduced three new people to Boswell, and these three people interacted with him at separate times across five days. Let's call them Good Guy, Bad Guy, and Neutral Guy. Good Guy complimented Boswell, was kind, offered him gum, and granted any request. Bad Guy was not complimentary, asked Boswell to complete tedious tasks, and refused any requests. Neutral Guy was nice but businesslike, not requesting anything of him, but not giving him anything either. Then Boswell was tested on the sixth day for his knowledge of these people. He could not remember or name any of the three people when shown their pictures. Next they showed Boswell a photo lineup of the three people together, plus a person he had never met. The researchers asked which of them he liked best, and Boswell consistently chose Good Guy above chance and Bad Guy below chance. Even more interesting, when measuring the amount of sweat he produced on his fingers, an automatic response, Boswell had a stronger physiological reaction to Good Guy than to any of the others. A part of his brain had implicit knowledge of Good Guy, even when Boswell could not tell the researchers anything about him.[4]

We have specific episodic memories of a loved one (a memory of our wedding day, for example), and the loved one is a part of many of our habits (how close we sit to them on the couch), but we also have implicit semantic knowledge about them (beliefs that they will always be there for us, that they are special to us). The implicit knowledge is stored in circuits of our brains distinct from where episodic memories are stored. This means we rely on different kinds of information about loved ones from different neural systems, which influence our thoughts, feelings, and behavior in their own distinct ways. When a loved one dies, over time and with experience we can refer to our episodic memories of their death—we know that they are no longer with us. But implicit knowledge is much harder to update, as it is responsible for the attachment-related beliefs that our loved one can be found, that we are not searching hard enough for them, that if we tried harder or were better in some way, they would return to us. Because this implicit knowledge conflicts with the episodic memories, we are less likely to acknowledge this implicit magical thinking. I call these conflicting streams of information the *gone-but-also-everlasting* theory, and I think it is because they conflict that grieving takes so long.

Episodic memory, habits, and implicit knowledge all influence how we understand, predict, and act in the world. While they may contradict each other (for example, episodic memory telling us that our loved one is gone, and implicit knowledge insisting they are not), all of them must get updated as we learn to live with their absence.

Why Does Grieving Take Time?

I can learn the names of all my students in a seminar in just a few weeks and gather information about their backgrounds. I develop a feeling for which student always has the answer; I recognize the ones who are funny or widely read, and I know those who do not volunteer to speak in class. I can even integrate this knowledge into our unfolding class discussions, asking simpler, fact-based questions to the shy students so they can give short, definitive answers, and more applied questions to those who are willing to think through their understanding out loud. This is a fair amount of information to encode about people, to remember, and to use. Yet all that information never adds up to the belief, the next semester, that any of those students will turn up in the classroom again. Grieving is different. Grieving takes more time. The *gone-but-also-everlasting* theory suggests that grieving is different from other kinds of learning, because the implicit belief in the persistence of our deceased loved one may actually interfere with learning about our new reality. In other words, episodic memory and habit, on the one hand, conflict with implicit magical thinking created through attachment, on the other hand, and this conflict leads to the extended period of time that grieving takes. I can easily understand that students from last semester will not be in my class today because there is no reason that they would be. But believing that my loved one is no longer on Earth, when part of how they are encoded in my brain as my loved one includes the information that they will be *here, now, and close*, takes time to understand and is not easy. Resolving incompatible beliefs interferes with learning.

If grieving were as simple as learning new information, creating new cause-and-effect predictions about the world, or making new habits for our day-to-day activities, I would not expect this learning to take months. It is true that any new knowledge requires time and experience to acquire, but the time it takes to acquire other types of knowledge compared to the length of time many people grieve suggests there is something else going on, like incompatible beliefs. Developing this new knowledge requires the willingness to engage fully in our life during bereavement, and we will talk more about engagement in our day-to-day life during loss in chapters 8 and 9.

Knowing We Have Magical Thoughts

Grief is the cost of loving someone. Bonding gives us the motivation to believe that when our spouse, children, and close friends leave us, it is temporary, and they will return. If we truly believed that they would not return every time they left for work or school in the morning, our life might be unbearable. Thankfully, we do not experience the death of our loved ones very often, in comparison to the number of times our loved ones come and go while alive.

When we lose a loved one, it is common to know that the person is gone and simultaneously harbor the magical belief that they will walk back in through the door again. If we take at face value that people believe both things, and accept that this is normal, then neuroscientists should look for multiple neural processes at work. We would want to see the perspective of the brain, where two distinct aspects of what they "know" can exist simultaneously. Considering multiple concurrent beliefs should give us a clearer picture of how the function of the brain affects the way we grieve. My own research has considered where in the brain these types of knowledge might reside, and in the next few chapters I will tell you more about how the brain overcomes these incompatible beliefs and restores us to a meaningful life.

CHAPTER 4

Adapting Across Time

When I was five years old, we had the electric heaters in our house replaced. I was not yet in school, and I became obsessed with our electrician, Jack. I followed him around, despite my mother's scolding. Jack always wore denim, and I, too, began to prefer my overalls. I can vividly remember his slow smile, the deep sense of kindness this larger-than-life man offered me. In a completely different experience of the adults in my small hometown, when I was in the fourth grade, I took art lessons from a local artist. I, and everyone else, called her by her last name. Weber was unlike anyone I had ever met, not least of all because she was the first woman I knew who did not shave her legs. Weber painted the most remarkable, detailed botanical watercolors of Montana wildflowers, two of which hang in my hallway to this day. Although I did not have any talent as a fine artist, I continued to visit and talk with Weber through high school and later, during my visits home from college at holidays and summers.

In what I perceived as a teenager to be one of the most unexpected relationships, Weber and Jack fell in love. They came to marriage relatively late in life, and were overjoyed when Weber became pregnant. During the pregnancy, however, Jack was tragically diagnosed with cancer, a devastating sarcoma. In one of the many attempts at any possible treatment, they came to Chicago, and I took care of their baby, Rio, one afternoon in my off-campus apartment while they went to doctors' offices. In a cruel and unfathomable twist of fate, Jack died when their son was only one and a half years old.

The paintings that followed when Weber was able to pick up a brush again were unlike her work before. Wildflowers still appeared in her paintings, but there were also clouds that dripped tears, women with tears that fell into buckets, and hearts from which endless teardrops of blood were wrenched. Many depicted women lying immobilized, covered by the leaves of wild raspberries or pinned by winter-bare trees. A huddled woman appears with heavy quilts laid over her, and in some, the black figure of grief wraps around her shoulders, burdening her like a heavy cloak. Yet, in the final paintings of the series, we see the woman retrieving her heart from its burial place underground, and in several, the sun finally appears, the first yellow-orange rays breathing light into the picture. These pieces are breathtaking.

In conversation with her one day in her studio, she told me that her artist's training had been invaluable in her grieving process.

Before, she had worked hard and developed great technical skill with brush and water and pigment. After Jack died, she really had something to say, and without those years of preparation, she would not have had the skill to convey the depth of her feeling. I could see that without the oceanic depth of feeling, her earlier work, while beautiful, did not evoke the same resonance in the viewer. A very long road stretched between Jack's death in 1996 to her gallery show in 2001, eventually restoring a new life, inspired by the presence of his absence.

How to Take a Picture of the Brain at Work

For many of us who have known grief, we resonate with Weber's paintings, overcome as the recognition of beautiful images and juxtapositions elicits our own grief experience. In the introduction, I began telling you how the first neuroimaging study of grief came about, when all the stars aligned for our project. Our question was, What happens in the brain when someone is experiencing a wave of grief—but how could we evoke the feeling of grief in the unfamiliar, sterile, medical setting of the neuroimaging scanner? The depictions that Weber created evoke the deep solitude and silence of grief; how could we reliably elicit that feeling? Scanners bang and whine loudly, and back then, people even had to put their teeth on a bite bar to keep their head from moving—not exactly a setting that allows people to access their deepest inner feelings.

Functional magnetic resonance imaging (fMRI) can pinpoint what part of the brain is active when a particular thought, feeling, or sensation happens. Neuroscientists infer where neurons are firing by looking at which brain regions have increased blood flow during this particular experience. FMRIs detect the blood flow because of the iron in blood, using the huge magnet that gives the technology its name. Then data from the blood pulsing through the brain is transformed through some complicated physics into the resulting images of brain activation. Neurons require blood after they fire, to bring restorative oxygen. Specific neurons fire when mental events happen, so we can see which brain regions are activated during these mental events based on where in the brain the blood flows. The regions of the brain that are significantly more active during the mental event of interest than during a control task are displayed as colorful blobs laid over a grayscale picture of the brain, with brighter colors representing more oxygen in the blood in a particular area used for that mental function. This is what people mean when they say the brain "lights up," but these colors represent the computed likelihood of activation in an area, and not actual light or color in the brain.

Most neuroimaging is based on the subtraction method. First you develop a task that requires the mental function that you are interested in, and you scan the brain of a person who is doing it. For example, let's say you are interested in the mental function of reading. The brain is active all the time, doing all sorts of things. While a person is reading, their brain is also experiencing physical sensations, keeping them breathing, recording what is happening into memory, and so on. In the subtraction method, the researchers come up with a second task, called a control task. The control task is the same in every way to the first task, except for the mental function the scientists are interested in. The brain is scanned while participants are doing both tasks sequentially. A control task for reading should account for the fact that the person is moving their eyes from left to right, across combinations of symbols that appear frequently in their native language. The control task might require people to look at nonsensical "words" that are made up of letters and syllables that are common in English but do not actually mean anything, so it is not possible to read them. For each brain scan, a computer records the brain regions active during the reading task and during the control task. When you subtract the activation during the control task from the activation during the reading task, the brain areas that are left are inferred to be the areas that are important for the mental function of reading.

To choose a task that could be used for evoking and studying grief through the subtraction method, Harald Gündel, Richard Lane, and I had to think about how to capture a brief emotional moment of grief. We considered how grief comes about in real life and settled on two possibilities. First, when people tell us the story of what happened to their loved one, the specific words that they choose are tied to their specific memories of the loss. Second, when a bereaved person wants to share something about their loved one, they often pull out a photo album. Words and photos were exactly what we asked each participant to share with us. Because grief is so unique, so specific to the beloved person who has died, we knew that we could not use the same words or photos for all eight women in the study. So, we digitized individual photos of the deceased loved ones that each participant brought us. In the digitized image, we added a caption using grief-related words that the participant had used in an interview about their loss. These were words like *cancer* or *collapse*, specific to the death of their loved one. During the neuroimaging scan, they looked at the photos and words while we measured their brain activity.

Next, we had to create the control condition. The brain has specific areas for identifying human faces and areas for reading words. We decided to use a photo of a stranger as a comparison. For matching the words, we used neutral words of the same length and the same part of speech. For example, the matching word for *cancer* was *ginger*. Thus, as our control task for the subtraction method, we made slides for each participant of strangers captioned with neutral words.

The photos that our generous participants shared were very moving—for example, from a woman who lost her husband of many decades, a photo of the handsome young groom with a piece of wedding cake. Another was a man in a Hawaiian shirt, his relaxed smile relaying through the camera the enjoyment of a shared vacation with the woman who was now his widow. When we asked the bereaved participants to tell us what they felt during the slide presentation, participants told us they felt the most grief when they looked at their loved one captioned with the grief-related word. We also measured the amount of sweat their fingers produced in response to each slide and they had the largest sweat response to those of the loved one with the grief word, and the smallest sweat response when they looked at the stranger and the neutral word, as we expected.

Usually in a laboratory study, we use the same stimuli for every person in the study, to hold that aspect constant. Asking bereaved people to bring in a photo of their loved one, so that every person looked at a different photo, was a novel idea. It was critically important, though, to evoke real grief in each person, because for each of us, our grief is as unique as our relationship.

Results

I mentioned the posterior cingulate cortex (PCC) in chapter 2, in the choose-your-own-adventure study. The PCC is a large region starting in the middle of the brain that curls around the central fluid-filled ventricles toward the back of the head. From other neuroimaging studies, we know that the PCC is important when retrieving emotional, autobiographical memories; in fact, the PCC enables the feeling of grief. Our reminders of the deceased loved one in the scanner sparked those memories for our participants. In our study, the PCC showed more neural activation when we compared looking at a photo of the deceased to looking at the photo of a stranger.

The PCC was not the only region activated during the grief task, however. A more contemporary understanding of brain function reveals that many regions are active at once, in a network. Another activated region is called the anterior cingulate cortex (ACC). Many mental activities require the ACC, because this region directs our attention to things deemed important. When we think about words that remind us of our loved one's death, compared to looking at neutral words, you can understand why that activates the ACC. Of course, the death of a loved one ranks high in importance—as a neuroscientist, this result reminds me of exactly how important it is.

We often see two regions, the ACC and the insula, activated together when something painful demands our attention, and we saw that co-activation during these moments of grief in the scanner. One reason we know so much about the ACC and insula engaged in joint activation comes from studies of physical pain. These two regions respond together during a physical pain stimulus, like uncomfortable heat applied to the participant's fingers during a neuroimaging scan. What is fascinating about the regions that are involved in physical pain is that neuroscientists can distinguish between the physical aspect of pain and the psychic, or emotional, aspect of pain. If you think about it, the physical aspect of pain equals intense sensation. Anatomists have long understood the neurons that snake through the body from sensation receptors on fingers, through the spinal cord, and into specific brain areas that have a topographical map of the body, indicating in consciousness where the pain sensation happened. But these neurons terminate in the sensorimotor brain region. So, physical pain derives from intense sensation produced in the brain. The emotional part of pain, the suffering that accompanies physical pain, derives from the ACC and insula, responding to the alarming and suffering aspect of pain. Thus, when these two regions were activated during grief, we interpreted their co-activation as related to the emotional pain of grief. The exact locations in ACC and insula are not identical in physical and emotional pain, but are very near neighbors.

Results Lead to More Questions

The results of this first study pointed out that grief is a very complex thing for the brain to produce. It requires many brain regions beyond those that process pictures and words: grief involves brain regions that process emotions, take the perspective of another person, recall episodic memories, perceive familiar faces, regulate the heart, and coordinate all of the above functions. On the other hand, the results were specific, confirming that grief does not activate all regions of the brain. For example, in our study grief did not activate the amygdala, an almond-shaped piece of the brain that is often elicited when the brain is producing strong emotions.

Our neuroimaging study demonstrated that grief could be successfully examined in the brain, demonstrating what occurred when we peered in. It was an important step forward for science to consider investigating grief from the perspective of the brain. On the other hand, the results felt incomplete, as they are just a description of regions involved. The results do not answer some of the important questions people want to know about grief. We needed a neurobiological model of grieving that goes beyond a laundry list of brain regions.

Back then, I believed neuroscience could provide insight into how the experience of grief changes over the period of grieving, in other words, how the knowledge of our loved one's absence is updated over time. I hoped that neuroscience could help us understand and predict who adjusts resiliently following the death of a loved one and who struggles to restore a meaningful life. In addition, I wanted to know how the brain might get in the way of our adapting. But these were early days when the first neuroimaging study of grief was published in 2003. This study of grief had created a foundation to describe what the brain did in the moment you feel grief, but it did not satisfy my scientific curiosity about the process of grieving.

Sharing Science with the Public

Simple description of a phenomenon is common in the early days of studying it, an initial step in training our focus on a new area of inquiry. One very famous description of grief has persisted in our culture for decades. In 1969, Elisabeth Kübler-Ross published *On Death and Dying*. The model of the five stages of grief that Kübler-Ross discussed in her book is the model the world remembers, despite the fact that research progress in the decades since has shown that model to be inaccurate or incomplete. This widespread awareness of Kübler-Ross's model is partly because she touched the hearts and minds of those who read her popular book. Everyone knows these stages (denial, anger, bargaining, depression, and acceptance), whether you wrote them on index cards to study for your Psych 101 class, or you merely googled *coping with grief*. That said, the information that you can find on the Internet about grief has improved somewhat, especially if you look at websites produced by good sources like the National Institutes of Health.

Elisabeth Kübler-Ross was a fascinating person. (I had the honor to hear her speak in Arizona, where she lived before her death in 2004.) She grew up in Zurich,[1] and as a young person, volunteered to work with refugees after World War II. She visited the concentration camp near Lublin, Poland, and the experience had a life-long, profound effect on her. In the 1960s, as a psychiatrist in the United States, she started seeing patients and writing during the civil rights and women's rights movements. These cultural shifts gave a voice to groups who had previously been voiceless. She similarly gave a voice, through her writing, to terminally ill people. The belief then, and to some degree even to this day, was that impending death is not something to be discussed, not even between doctor and patient. She chose instead to interview patients about their experiences of tremendous loss as they faced their mortality, asking what they felt, what they thought about, and how they understood what was happening to them. Not only that, she invited other nurses, doctors, residents, chaplains, and medical students to join these interviews. Then she shared what these real people who were dying had to say, first in a feature article in *LIFE* magazine, complete with moving photographs of these interviews, and then in her remarkable book in 1969.

Kübler-Ross was using some of the best technology that psychiatry had to offer back then—the clinical interview. She did what all scientists do when they first begin to study a phenomenon: she described. She cataloged what patients said, and she distilled what they described into a model and shared that model with the world. She was not wrong about the content of grief. People described experiencing anger and depression. Some of them could not report on their experience because of denial, and others spent a great deal of time and effort ruminating on how they could bargain their way out of death. Some seemed at peace with what was to come, accepting that they were in the last chapter of life. She described what they shared, focusing on and creating a model that included those aspects that seemed most important, in a way that no one else had done.

Kübler-Ross and others applied the stages of grief she originally described in terminally ill patients to grief in the wake of bereavement, which is a big leap. But description is not the same as empirical investigation. Just like with my first neuroimaging study, there was more to find out about grief. Kübler-Ross was using people's momentary experience of grief during interviews to describe grieving over time. Although she was correct about reporting the content of people's experience, not all people who are grieving go through all of the five stages or go through them in that order. The five stages are not an empirically proven model of the process of adaptation after loss.

The problem, and the damage that this has caused for bereaved people, is that the model she developed has been considered more than a *description* of grief of those she interviewed, and taken as a *prescription* for how to grieve. Many bereaved people do not experience anger, for example, and therefore feel they are grieving wrong, or have not completed all their "grief work." Clinicians may say that a patient is in denial, without understanding that the stages are not linear, and that people go in and out of denial at different times. In sum, very few people experience the orderly progression of stages that Kübler-Ross proposed, and tragically, they may feel they are not normal if they do not. This old, outdated model has been replaced with models that have more empirical science behind them, but clinicians sometimes persist in using it, and the general public is usually not aware that our understanding of grieving has developed significantly.

The Hero's Journey

When I tell people that I am writing a popular science book about grief, almost everyone I speak to assumes that I will be discussing the five stages of grief. Why does this model persist despite scientific evidence that grief does not proceed in linear stages? Psychologists and grief experts Jason Holland and Robert Neimeyer have proposed the best reason for this persistence I have come across.[2] They describe the five stages model as reflecting our culture's "monomyth." The hero's journey, or in this case, the griever's journey, is an epic narrative structure we find in most books, movies, and campfire stories we have ever heard. You can think of any hero from Ulysses in the *Odyssey* to Alice in *Alice in Wonderland* to Eleven in *Stranger Things*. The hero (griever) enters an unfamiliar and terrifying world, and after an arduous journey, returns transformed, with new wisdom. The journey is composed of a series of nearly impossible obstacles (stages) to be overcome, making the hero noble when they succeed in their quest. Holland and Neimeyer put this well: "the seemingly magnetic draw of a stage-like depiction of grieving that begins with a disorienting separation from the 'normal,' pre-bereavement world, and that progresses heroically through a series of clearly marked emotional trials before eventuating in a triumphant stage of acceptance, recovery, or symbolic return, may owe more to its compelling coherence with a seemingly universal narrative structure than to its objective accuracy." The problem with this monomyth is that people feel they are not normal when they do not experience a linear set of obstacles. Or they feel like failures because they have not "overcome" grief or achieved

74

some enlightened state. Friends, family members, and even doctors may worry when there is no clear return of a wise hero.

Holland and Neimeyer conducted an empirical study that looked for the five stages and found that adaptation is not so linear or orderly. Grief distress is usually more pronounced in folks who have been grieving for a shorter period of time. But the distress includes all types of grief experiences, including disbelief, anger, depressive mood, and yearning. Acceptance is most evident among those who have been grieving for a longer period of time. Thus, grief distress and acceptance seem to be two sides of a coin, but the rise and fall of each one tends to look like waves across days, weeks, and months. The relative increase in acceptance as compared to the relative decline of grief distress does happen, thankfully, but over a long period of time. In the midst of this slow inversion of acceptance over distress, there tends to be a temporary reversal around each anniversary of the death, when many people experience a normal recurrence of their grief. The journey doesn't typically have a clear beginning, middle, and end that we may hope for, or that our loved ones may hope for us, in the midst of our distress. On the waves of grief, eventually acceptance rises more often, and distress falls off in intensity without completely disappearing.

The Dual Process Model of Coping with Bereavement

Bereavement science moved slowly in the late twentieth century from concentrating on the content of grief that people experienced to focusing on the process of grieving the loss over time. Through a long collaboration, psychologists Margaret Stroebe and Henk Schut at the University of Utrecht in the Netherlands provided elegant empirical bereavement science and developed a model that many clinicians use now, the dual process model of coping with bereavement, usually just called the dual process model for short.

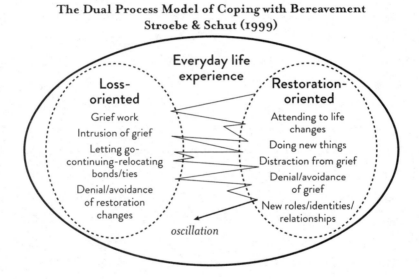

The Dual Process Model of Coping with Bereavement
Stroebe & Schut (1999)

Everyday life experience

Loss-oriented
Grief work
Intrusion of grief
Letting go-continuing-relocating bonds/ties
Denial/avoidance of restoration changes

Restoration-oriented
Attending to life changes
Doing new things
Distraction from grief
Denial/avoidance of grief
New roles/identities/relationships

oscillation

Take a look at the picture of the dual process model. The outermost sphere represents our everyday experience, as we go about our day-to-day life. The two ovals inside represent the stresses that we face when a loved one dies. For decades, clinicians, philosophers, and poets have been talking about loss-oriented stressors—the painful emotions of losing someone, the way that everything seems to remind us of them, even though we know they are gone. These stressors constitute what we typically think of as grief. The important addition of the dual process model was to name the other stressors that we face. For example, we also face what Stroebe and Schut called restoration-oriented stressors. These are all the tasks that we now have to do because the person is gone. Restoration stressors include practical things that you are not used to doing, or at least not doing alone, such as figuring out your taxes or shopping for groceries. In the case of losing a spouse, you not only have to learn to live without your friend and lover, but also without the person who used to do housework, say, or without a co-parent. For an older couple, widowhood might mean living without a significant support for our health issues, or without the person who always did the driving. And restoration means reorienting to how our world has changed, such as recognizing that our retirement dreams are not going to happen with our loved one. We have to make new choices and develop new goals in the face of our new reality in restoring a meaningful life.

The real genius of the dual process model, however, is the jagged line connecting the two ovals in the figure, showing that people go back and forth between these stressors. This oscillation line highlights the process of grieving, rather than only the content of our thoughts and feelings. Sometimes the oscillation occurs within one day; for example, you tour houses with a real estate agent in the morning and become absorbed in your wedding album memories in the afternoon. Sometimes it is even shorter, like having a cry in the office bathroom and then returning ten minutes later to the project on your desk. Sometimes facing one stressor means complete denial or avoidance of another one: "I am just going to pretend that nothing is wrong for the next forty-five minutes and cheer for my daughter's soccer match."

When the seeds of the new dual process model first sprouted, some clinicians challenged it, because the model punctured some closely held beliefs (or myths) about grieving—for example, the myth that grieving requires us to focus solely on confronting the feelings of grief, without any consideration for the fact that the grieving person may also benefit from time spent not confronting those feelings. Time off from grieving might look like denying, suppressing, or distracting oneself from one's feelings about the death, and this was presumed to be bad for long-term adjustment. But time off from grieving can give your mind and your body a break from the stress of the emotional upheaval. Stroebe and Schut wanted to address these limitations in the previous models of grieving.

Both poles, addressing loss as well as restoration, are important to the experience of grieving. The key to coping well after you lose someone is flexibility, attending to what is happening day-to-day, and also being able to focus on coping with whichever stressor has currently reared its ugly head. Bereaved people also have times when they are not consumed by grief, when they are simply engaged in everyday experience outside the two ovals. As time passes, they are more and more engaged in everyday life, and the difficulties of loss and of restoring a meaningful life recede gradually. The ovals representing loss disruption and striving for restoration never disappear, but these stressors evoke less intense and less frequent emotional reactions. In the second half of the book I will discuss in greater detail how this flexible approach to coping with loss works.

 CHAPTER 5

Developing Complications

In the summer of 2001, I was invited to attend a workshop at the University of Michigan, just weeks after collecting the first neuroimaging scans of grief. Leading grief researchers from the US and Europe attended, and the workshop had an enormous impact on me, expanding my understanding of how to think scientifically about bereavement. That weekend I met wonderful people and scientists, including George Bonanno, Robert Niemeyer, and Margaret Stroebe, who have brought bereavement science into the twenty-first century. They encouraged me in my work as a young researcher and have continued to influence me as we have developed into colleagues over the years.

The purpose of the workshop was to introduce us to the Changing Lives of Older Couples (CLOC) research project, done at the University of Michigan and funded by the National Institute on Aging. This project has greatly influenced the field of bereavement research. In this longitudinal study, more than 1,500 older adults were interviewed, with hundreds of questions, across different time points before and after the death of a spouse. As you might imagine, that creates an enormous database. The workshop showed us what information had been collected, how it had been compiled, and what research questions had been answered so far. More than fifty scientific papers, several of which have been groundbreaking, have come from this research project to date.

One of the most valuable things about the CLOC study is that the participants were first interviewed when both members of the couple were still alive. When the original interviews were done, neither of the spouses was terminally ill. Then the researchers kept track of these couples for many years. When one of the spouses in the couple passed away, the surviving spouse was interviewed again, at six and eighteen months after the death. Because in the original interview there was no indication of when a member of the couple would die, this is a unique type of study, a "prospective" study. Information came from the couples prior to widowhood, so we are not relying on the widow or widower to recall what things were like before their loss. Having prospective information prevents inaccuracies, since our memories are affected by time and biased by events that have happened in the intervening period.

The perspective from before the death of a spouse has proven invaluable in empirically debunking some of the myths about grief. From this CLOC data, George Bonanno developed an empirically-supported model of grieving using the information about changes in grieving over time, and his model of these adaptation trajectories influenced the field enormously. Imagine how different Kübler-Ross's model might have been if she had lived in the era of science with access to 1,500 bereaved people and interviews at multiple time points across years! Data sets of this magnitude reassure us that the patterns of adaptation are reliable across a wide number of people. Many interview questions all in one database have allowed scientists to test the associations and even predictions between the emotional, personal, circumstantial, familial, and social aspects of grieving.

Trajectories of Grieving

Imagine that you join a book club. At the first meet-up, you are introduced to a woman who tells you she was widowed about six months ago. You notice she seems withdrawn and at the same time restless. She is the first one to leave that evening. You hope she will return, as she seems nice and has some interesting thoughts about the book. Indeed, she attends the group every month. Sometimes she seems a little better and sometimes a little worse, but basically about the same. The book club is enjoyable and you keep going until eventually you realize that you have been attending for about a year and a half. This strikes you because you realize not much has changed for this woman during that time. She does not talk about any new people in her life, she often gets teared up when the book has any type of loss in it, and she just seems, well, depressed.

Hold her in mind while we jump back into scientific models.

The insightful question that Bonanno answered with the CLOC data was this: Does everyone's adaptation trajectory during griev- ing look the same?[1] If bereaved people were interviewed at six months and eighteen months after their loss, would everyone look the same, or would you be able to detect groups of people who fall into different patterns? In fact, in the CLOC study Bonanno and his colleagues found that there were four trajectories that could be used to categorize people's grieving. These trajectories include *resilient* (those who never develop depression after the death of a loved one), *chronic grieving* (depression that begins after the death of a loved one and is prolonged), *chronic depression* (depression that began before the death of a loved one and continues or worsens after the death), and *depressed improved* (preexisting depression that abates after the death of a loved one). This model of the trajectories of grieving has now been replicated in several other large studies. It was nothing short of remarkable to have such fine-grained data about the grieving process of so many individuals.

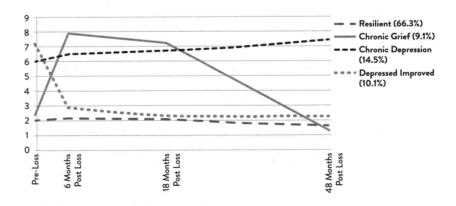

Let's consider which trajectory fits the woman from the book club. In the figure, the numbers on the vertical *y* axis (on the left side) indicate depressive symptoms; higher numbers represent higher levels of depression. The woman from the book club was depressed at six months after her husband's death when we met her, and still has depression at eighteen months. But here is the real insight of the trajectories in the grieving model. You do not know if that woman falls into the chronic depression group or the chronic grieving group. That's because you met her after her husband died. The difference between these two trajectories is whatever was happening in her life before his death.

If she fits into the chronic depression group, this woman was struggling with depression before he died, and bereavement is a continuation of the difficulties she was experiencing. If she fits into the chronic grieving group, she was getting along in life, with normal ups and downs, but not suffering from depression. It was the death of her husband and the stress of his ongoing absence that led to her depression. Once she was depressed, she could not climb out of it for months and months. You can probably imagine why the difference between these two trajectories is important. In one case, her struggles are long-standing and probably require a different type of intervention than if her problems started with widowhood. Bonanno's insight could only be shown with prospective data. When a clinician is faced with a person who is suffering during bereavement, they need to ask if it is a long-standing problem. We must not assume that the death can be pinpointed as the cause of the suffering, even though they are suffering after the loss.

You may notice that by four years, or forty-eight months, the woman experiencing chronic grieving has the same level of depressive symptoms as those who followed a resilient trajectory. We know that there are people who experience chronic grieving for much longer, even a decade. So, even in the trajectory of chronic grieving, adaptation is possible, even if the process is much slower.

Resilience

One of Bonanno's trajectories of grieving was "resilience." These widows and widowers did not have depression before they lost their spouse, and when they were interviewed at six months after losing their spouse, they still did not show signs of depression. The same thing happened at eighteen months. Of course, we cannot say what they felt in that first six months, and just because they did not have depression does not mean that they did not experience grief or distress.

What was remarkable, however, was how many fell into this "not depressed" resilient category: more than half of the widowed spouses. That means that resilience is the most typical pattern of grieving, showing that most people who experience the death of a loved one do not experience depression at any time point. Frankly, this surprised many people who study grief. This insight reminded us that clinicians had primarily been studying bereaved people who sought help after their loss, a smaller group than the larger "resilient" group who didn't experience depression. We had generalized our understanding of the people who were having difficulty coping to extend to all bereaved people, because we didn't have systematic, large-scale research on grieving. We only achieved this knowledge about the common experience of resilience because the CLOC study had randomly chosen people in Detroit to participate in the study. Random sampling requires careful social scientific methods and is harder than you might think. When subjects were first asked to join the study, the researchers did not know how they would cope with widowhood, because they had not yet lost a spouse. That meant that people who adapted well and people who did not adapt well were equally likely to be included.

Interestingly, there is less research about grief that is not very disruptive in people's lives. For clinical psychology, this makes sense, because the clinical motivation is to understand what helps people who need help. It is also easier to get people to volunteer for a study when they are looking for help. But it can skew our understanding of what bereavement is like.

Grief versus Depression

Sigmund Freud was the first to write about how similar grief and depression are.[2] Though they can look the same, one difference between them is that depression often seems to come out of nowhere, whereas grief is a natural response to a loss. Since the time of Freud, we have learned that depression and grief, even severe grief, can be distinguished. For example, depression tends to pervade every aspect of life. People who have depression feel that almost all facets of their life are awful, rather than feeling that it is just the loss they are struggling with.

My mom died when I was twenty-six, and I did not develop complicated grief, but I struggled with depression. As I mentioned earlier, my mom also had significant depression, starting with episodes before I was born, and she endured them through my child-hood. Depression runs strong in my maternal family, not unlike a vein of metal ore coursing through the generations, picking out one individual or another. I had already experienced one episode of depression before she died, during a period of homesickness in my junior year abroad in college. My response to her death included another bout of depression, and it was not my last. As I learned more from people who experienced complicated grief in my re-search studies, I came to realize that the hallmark of their grief experience was yearning. That was not the feeling I wrestled with when I was grieving. Although I struggled after the loss of my mom, I was not yearning for her to be around again. If anything, I felt relieved that she was gone, because my relationship with her had been so difficult, and because I knew how unhappy she was during stretches of her life. Feeling relief over a loved one's death, though not uncommon, is terribly stigmatizing, so I did not admit that to a lot of people. In fact, I still have trouble admitting it to you now. Without her in my life, there was less interpersonal conflict, but many of the relationship patterns I developed through two de-cades with my mom were repeated in my other relationships, and so my depression pervaded many aspects of life.

Unlike my situation, for a person with chronic grieving, the awful feelings stem from missing the deceased, and if there is guilt, it is also focused on something about the loss. In other words, if the deceased loved one were once again alive, the person with depres-sion might be glad, but the return of the loved one would not solve everything. They would still be depressed. But for a person with chronic grieving, the feelings, the distress, the difficulties are all tied to the absence of the person who has died. Anecdotally, people who have experienced depression earlier in their lives say that grief feels different from depression.

Bereavement science recognized that there were people who began struggling after the death of their loved one, and continued to struggle for months and even years. A group of experts on grief and trauma, including researchers and clinicians, convened in 1997 to discuss whether they could agree on the symptoms of a chronic grieving disorder.[3] Although many people had written about those who do not recover after a loss, there was no clinical consensus on what criteria should be used to identify this chronic grieving phenomenon.

This group of experts identified a list of symptoms characterizing those who were having the most trouble adapting after the death of a loved one. They agreed, based on empirical evidence and clinical experience, that a grieving disorder could be distinguished from disorders of depression or anxiety (including post-traumatic stress disorder). The primary symptoms of this chronic grieving included (1) preoccupation with yearning for the deceased, and (2) traumatic symptoms caused by the loss. Criteria were developed that clinicians and researchers could use to determine if a person they were studying fit this phenomenon of chronic grieving. Creating these criteria was important, because previously, different researchers had used different definitions of what constituted severe grief, making it difficult to compare research studies.

By clarifying the set of symptoms of a grieving disorder, we could start to ask other scientific questions. For example, we might be able to predict, and support, those who were at higher risk.

We could ask whether there were other features associated with chronic grieving, like physiological stress or the way loss was processed in the brain.

Prolonged Grief Disorder

There is an upside and a downside to calling chronic grieving a disorder, thus giving a name to an experience that afflicts a small proportion of bereaved people who struggle very intensely for a very long time. The upside is that naming a disorder lets people know that others have struggled in the same way, which can be very reassuring. It lets them know they are not the only ones, and that researchers are actively working on how to intervene. Although developing clinical criteria is not my primary area of study as a clinical scientist, understanding the neurobiology of grief is very difficult without some background in this diagnostic history. We cannot comprehend what might go wrong in the brain during chronic grieving without understanding what might go wrong psychologically.

Once we understood that one in ten bereaved people do not adjust over a long period of time, we focused our clinical attention on those who did not improve with the typical support of their friends and family. This small portion of people do not return to feeling their lives are meaningful over time. Focusing on those who have a grief disorder, by using these criteria, has led to psychotherapies that can alleviate this disorder effectively. I will talk more about these treatments later in this book.

We, as scientists and clinicians, are still in the early days of understanding exactly what this grief disorder is. We are still in the process of distinguishing it from the normal human suffering of grief, and distinguishing it from depression, anxiety, and trauma. Because we are still in the middle of making history, disordered grieving has developed a few different names, including *complicated grief* and *prolonged grief disorder.* Though initially used by the group in 1997, the term *traumatic grief* has come to mean the grief after a traumatic death; the term *traumatic* focuses the emphasis on surviving a death that is sudden or violent. Prolonged grief disorder is now included in the International Classification of Diseases (ICD-11) produced by the World Health Organization. It was accepted as a diagnosis in the *Diagnostic and Statistical Manual of Mental Disorders* (DSM-5-TR) produced by the American Psychiatric Association in 2022. Characteristic symptoms include intense yearning, or preoccupying thoughts of the deceased, on a daily basis. Among other symptoms, there is intense emotional pain, a feeling of disbelief or an inability to accept the loss, difficulty engaging in activities or making plans, and a feeling that part of oneself has been lost. These symptoms occur for at least six months (or for at least a year, in the DSM-5-TR), interfere with the ability to fulfill one's job, school, or family responsibilities, and exceed what is expected in the person's cultural or social context.

The lives of this small group of people with a grief disorder are different from those who experience the universal human suffering of grief. I see it in the woman who told me there was no reason to give her children bar mitzvahs if their grandmother was not there for them. I see it in the man who was a leader in his local community, but after the death of his son, could no longer be a resource because he "just didn't care about people anymore." I see it in the reporter for a national newspaper who eventually lost her job because she could not get through an interview with her sources without crying. This is the experience of the widow who continues to buy the same amount of groceries as she did before her husband's death, despite knowing she will throw out the uneaten half of the meals she cooks for two.

I like the term *complicated grief*, because it reminds me of complications that can happen in any normal healing process. If you break a bone, the body creates new cells that remodel the bone and return it to its original strength. Although doctors might support this process by stabilizing the bone with a cast, knitting bone back together is a natural healing process. Even years later, if you have broken a bone, from an X-ray a doctor can still tell it was broken. Grief is similar, in that anyone's life is forever changed because of loss, even when they have adjusted well. However, there can be complications with a healing bone fracture, like an infection or a second injury, and I think of prolonged and severe grieving in the same way. Usually there are complications that have interfered with the common adaptation process, and the goal is to identify and solve these complications in order to get a person back on track with typical, resilient adjustment. Later on, we will look in depth at one type of complication, created through certain thoughts that arise as we are adapting.

In this book, I most often use the term *complicated grief*, the term that was in vogue when most of the research I report on was conducted. I am referring to the severe, prolonged experience that results from complications in grieving after a death. This is "chronic" grieving, the upper end of the continuum of grieving that can be called a grief disorder. In current clinical science, complicated grief captures a larger number of people in that upper end of the continuum (about 1 or 2 in 10) than does prolonged grief disorder (between 1 and 10 in 100). Although the terms are somewhat different, my point is primarily to indicate bereaved people who fall in the upper end of this continuum.

Grief and the Structure of the Brain

Are there differences in the brains of those who are adapting resiliently and those with complicated grief? The death of a loved one affects the brain, but the relationship between grief and the brain is a two-way street. Brain function, which depends on the structural integrity of the brain, also affects our ability to understand and process the death event and what it means for our life. To put it most dramatically, if a person cannot remember well, or cannot form new memories, they have to be told and retold over and over again that their loved one has died. Without the brain structure to hold the memory in place, they are freshly confronted with the loss again and again.

Our cognitive capacity to keep memories, make plans, remember who we are, and imagine the future may help us restore a meaningful life. Science has investigated how the brain function and structure of the bereaved person impact the relationship between these mental capacities and grief outcomes. Researchers at Erasmus Medical Center in Rotterdam have published a series of studies that shed light on how our cognitive processes and our brain change during bereavement. In 2018, I had the good fortune to work with these researchers when I was on sabbatical in the Netherlands.

Back in the mid-1980s, these prescient doctors and researchers realized older adults would become a greater portion of the Netherlands' population, just as we face the graying of the United States. They knew that this demographic shift would cause a rise in older adults with chronic diseases, and that the best way to discover the causes of these diseases was to study the risk factors. They began a huge epidemiological study.

As I have discussed, teasing apart the causal aspects of disease requires prospective research. People must be assessed before they develop a disease, and then they can be followed to pinpoint when they develop heart disease, cancer, or depression. With this before-and-after information, researchers can look back and see what causal factors existed. Significantly, because of the wide range of people sampled, they can also look back at whether these factors existed for those who did not develop these same illnesses.

The Dutch researchers had the brilliant idea of focusing on one typical neighborhood in Rotterdam, and they built a special medical research facility in the middle of that district. This allowed regular medical and psychiatric assessments, central record-keeping, and an integration of the community and the researchers. For research on grief, they made a key decision that would dramatically change bereavement science. Not only did they ask people whether they had experienced the death of a loved one, but they also asked about the standardized diagnostic criteria to assess the severity of their grief. Consequently, we now have years of information on the trajectory of many grieving older adults.

The Dutch people in the study also had structural MRIs of their brain. Structural MRIs are different from functional MRIs (fMRIs). Because it tells us where neurons are firing, I used an fMRI for the first grief study, to determine what parts are used for specific mental functions, like memory or emotion. Structural MRIs, on the other hand, distinguish bone, cerebrospinal fluid, and gray matter. A structural MRI is basically a fancy, three-dimensional X-ray. Structural MRIs can also be used to look at the knee or the heart. When focused on the head, a structural MRI shows researchers the overall size of the brain. Significantly, it also shows the structural integrity of the brain's gray matter and white matter. As it so happens, the brain is not solid. Instead, there are tiny spaces between all the neurons. Just as two bones can be the same overall size, if one bone has osteoporosis, it may be porous and brittle because of lots of extra holes inside, meaning its structural integrity is poor. So, the two bones can have the same size, but not the same volume. Similarly, in the brain, spaces are created as neurons shrink due to normal aging, injury, or disease. These can be detected with a structural MRI, and we can compare the brain volumes of different people.

The Rotterdam Study compared the brains of 150 older people with complicated grief, 615 bereaved people who did not have complicated grief, and 4,731 nonbereaved people. People with current major depressive disorder were not included, so the results were clearly associated with grief and not with depression. The group with complicated grief had significantly less brain volume than the nonbereaved group,[4] but the brains of the nonbereaved and the resilient bereaved groups were indistinguishable. So greater grief severity in older adults, and not just experiencing bereavement, was associated with a slightly smaller brain volume.

A single MRI scan is a snapshot in time, a cross-section of information. It cannot tell us anything about whether smaller brain volume is the cause or consequence of bereavement. Smaller brain volume in those with complicated grief does not shed any light on whether the structural differences existed before bereavement or developed afterward. On one hand, less structural integrity preexisting in the brain might prevent resilient adaptation to bereavement. On the other hand, the stress of severe grief might lead to a small amount of shrinkage in the brain. A slightly smaller, less healthy brain might make it harder for us to learn, or adapt, during grieving. The important point is that in a very large study of older adults, on average, some structural brain differences existed for those with the most difficulty adapting.

This finding begs the question of whether there are also changes in the cognitive functioning of people who are bereaved, or in those with complicated grief. Grieving is very mentally demanding. The mental capacity to plan for the future after the death of a loved one requires us to draw on our past experiences, generate and anticipate possible outcomes, and keep our larger values, goals, and desires in mind—all while considering our present circumstances and our general knowledge of the world. Integrating all of this information into a coherent plan that we can act on requires quite a lot of cognitive capacity!

Notably, many bereaved people complain of difficulty concentrating. Standardized cognitive testing can be done to determine whether bereaved people differ in their cognitive capacities from nonbereaved people. A bereaved person could have difficulty concentrating because of something other than one's cognitive capacity. For example, this lack of attention might be caused by background thoughts about the deceased or the loss. In contrast, if a bereaved group could not perform well on a cognitive test, even when they were giving it their full effort and attention, we could conclude that cognitive impairment was the cause of the difficulty. Fortunately, the same Rotterdam Study researchers who investigated brain structure also gave cognitive tests to their participants.

Cognitive Function in Bereavement, Now and Later

In the Rotterdam Study, the older participants underwent a whole battery of cognitive testing. This included testing short-term and long-term memory, information processing speed, attention and concentration, memory for words and their associations, and global cognitive functioning. These tests included, among others, word puzzles, matching symbols, recalling stories, and making patterns with blocks, all standardized for the person's age and educational background. Psychiatrist and epidemiologist Henning Tiemeier found that the resilient bereaved group did not perform any worse on these tests than the age-matched group who were not bereaved. Thus, bereavement alone does not affect cognitive capacity.

On the other hand, the group with complicated grief did not perform as well on the cognitive tests in comparison to the bereaved group who were more resilient. Those with complicated grief had slightly lower overall cognitive functioning and poorer information processing speed. Again, we know nothing about which came first; it's the-chicken-or-the-egg problem. Did the stress of adapting to a death affect cognitive functioning, or did the older person's cognitive functioning affect their ability to process the death and its aftermath? Poorer overall cognitive function may lead to more severe grief because it is harder to adjust to loss with less cognitive capacity. Alternatively, cognitive function may be impaired because a prolonged grief reaction may affect the structure or function of neurons, and consequently, the mental functions our brain enables.

There is a little bit of evidence that helps us tease this problem apart, although I do not think it is definitive. When the same older participants had cognitive testing seven years later, those with complicated grief were still more likely to have some overall cognitive impairment compared to the ones who were grieving resiliently.[5] The brains of the resilient bereaved people still looked like those who were nonbereaved. This data suggests that loss is a normal life event to which the majority of people adjust without lasting deficits. However, for those with complicated grief, something unique occurs. Tiemeier and his colleagues interpreted these results as follows: at least for older adults, people with mild cognitive impairments are more likely to have more severe grief reactions when their loved ones die. This mild cognitive impairment makes them more vulnerable to suffering from complicated grief.

The slow cognitive decline they are experiencing may happen over decades. One possibility is that the poorer cognitive functioning may not be caused by bereavement, but rather that cognitive decline is being attributed to bereavement because the event is easy to point to, even if the loss happened in the middle of a slow cognitive decline. I believe we still need more research in this area. I wonder whether, for these older adults with complicated grief, effective therapy that helped them to adapt better could slow or stop the cognitive decline.

It is important to note that there are some limitations with this research. For example, cognitive decline as an explanation for complicated grief reactions is less likely for those who are middle-aged or younger at the time of loss. Research studies with cognitive testing and structural MRIs have not yet been done in younger folks. Research also uses group averages. For any one person who develops complicated grief, we cannot say it was caused by mild cognitive impairment. Even if mild cognitive deficits are a risk factor for complicated grief, the decline over time is very likely to be an interaction of the aging brain and the stressful bereavement event.

In addition, psychotherapy for complicated grief may improve cognitive functioning. Australian clinical psychologists Richard Bryant and Fiona Maccallum used cognitive behavioral therapy (CBT) to treat a small number of people with prolonged grief disorder. Then they tested their ability to recall specific memories before and after treatment.[6] Psychotherapy enabled bereaved people to recall more specific autobiographical memories. Those who showed the greatest improvements in their grief during therapy also showed the greatest increase in this capacity for memory. Therefore, prolonged grief and poorer cognitive function may be associated, though not causally. If the prolonged grief remits, then the cognitive difficulties may resolve as well.

Psychotherapy for Complicated Grief

Imagine yourself at the checkout of a grocery store, buying food for the week. You watch the items go by on the conveyor belt and hear the beeps as the cashier passes the scanner over them. A widowed woman named Vivian found herself in this place, week after week. As she watched the checkout process, she thought to herself, "I know I'm just going to throw half of this away." Why? Because she was still cooking for herself and her late husband every night. She prepared the elaborate meals exactly the way she had always prepared them. Unable to eat for two, night after night, she would scrape half the meal into the garbage. And yet, the next week, she would find herself selecting the same number of vegetables, pasta, hamburger buns, and cartons of milk as the week before. She just could not bring herself not to shop for him, as though her unwillingness to feed him would cut the final thread in the sturdy rope that had bound them together for forty years. Unable to control anything else, she was still able to cook for him. At the same time, she knew that her actions made no sense. She did not put out a plate for him and serve his portion—there was no misunderstanding in her mind that he had died. But, because she feared her family and friends would think she was crazy, she hid this nightly routine from everyone.

Vivian eventually heard about Complicated Grief Treatment (CGT). Without a lot of hope, but with a glimmer of recognition that her months of uneaten dinners might fit the disorder the ad described, she made a therapy appointment. CGT was developed by psychiatrist Kathy Shear at Columbia University. Shear's randomized clinical trials proved that people can recover when therapy is targeted specifically at complicated grief symptoms, and more recover with CGT than a control group receiving another type of psychotherapy. Shear's studies have been published in the *Journal of the American Medical Association* (*JAMA*) and the *American Journal of Psychiatry.* Even in older adults, 70 percent who received CGT therapy recovered, compared to 32 percent who received another therapy.[7]

Vivian began the intensive sixteen-week therapy. Initial sessions focused on explaining how grief works, and her therapist communicated that many people feel that being stuck in grief is their own fault. Vivian definitely felt that way, and recounted that her extended family felt she needed to "move on." But the therapist talked about how together they would pinpoint the complications getting in her way, and he told her she would have homework between sessions to build different skills she needed in her life now. He taught her to observe and write down her thoughts and feelings, so they could discover which ones were most problematic for her.

The grocery shopping was the most obvious problem that Vivian could point to. The therapist said that this was one of the restoration stressors from the dual process model—how to manage grocery shopping and cooking. But he also wanted to focus on the loss, and asked if he could record her telling him how her husband died (Vivian had not really described the events of that day to anyone before). She explained that her husband had been in the hospital for a couple of weeks, and she had been by his bedside day and night. They were very close, and she wanted to be there on the few occasions he woke up. One afternoon, the nurse who had seen her there every day gently suggested she should go home, shower, and get some clean clothes to bring back. Vivian was exhausted, and so she agreed. An hour later, when she returned, the nurse told her that her beloved husband had passed away. Vivian was so overcome with grief and guilt, she could barely say these words to the therapist. "I never admitted that it was my fault to anyone before," she said. "He died without me."

CGT addressed the stress of loss by revisiting these intense and overwhelming emotions again and again, and teaching skills to move flexibly in and out of these feelings. Together Vivian and her therapist realized she was avoiding this memory, and they practiced strategies for revisiting it. Vivian's therapist asked her to listen to the recording of herself telling the story every day, encouraging acceptance of the reality of her loss. This homework requires great self-compassion to face the suffering of grief, and part of this self-compassion entails "dosing" the feelings and also learning to set them aside; this is the oscillation we see in the dual process model.

To address restoration stressors, the therapist asked Vivian what it would be like to cook a meal just for one. "Frankly, I'd rather not eat," she said. "It's just too depressing to imagine a tiny potato in a pot, or on a plate. I'd just feel so lonely." What else could she do with the food? Vivian decided to go out, buy some disposable containers, and start freezing the leftovers. She knew she wouldn't eat them, but she said she could check at her church in case anyone else needed meals. In fact, the volunteer coordinator for visiting homebound parishioners said that home-cooked meals were highly in demand. Vivian didn't really see herself visiting lonely people in their homes, she told her therapist, but she said she could bring her frozen leftovers to the church to be distributed by others.

For many bereaved people who have been suffering for a long time, finding goals and activities with a therapist that elicit even small amounts of interest is a revelation. Before therapy ends, the therapist and bereaved person work on strengthening social connections, finding or improving relationships with kind or loving people who will be in their life afterward. For Vivian, even trying a new way of doing things as an experiment inched her forward on an upward spiral. The volunteer coordinator turned out to be a bouncy young woman who couldn't get enough of Vivian's stories about her life and her travels around the world with her husband. And she enjoyed Vivian's cooking, too!

CGT provides a therapist-guided imagined conversation with the deceased. During one of these conversations, when Vivian said out loud how much she had loved him, she was flooded with the feeling that he loved her, too. "I think he loved me too much to die while I was in the hospital room," she said. "Maybe it was a blessing that I left, so he could let go the way he needed to." The strength of her feelings of love made her realize that what still held them together wasn't her cooking, but rather a deep bond that could never disappear. Later, though Vivian still cooked for parishioners because she found it meaningful, she no longer did it out of a compulsive feeling that she needed to feed her husband.

There are still relatively few therapists trained in evidence-based psychotherapy for complicated grief. In addition to CGT, other forms of psychotherapy that have an empirical basis include exposure therapy and cognitive-behavioral[8] therapy. In Europe, studies have shown targeted cognitive-behavioral therapy can be effective in a group setting as well. But bereavement science is making strides in understanding what the key ingredients for therapy are for those with complicated grief, and what has to change for the bereaved person in therapy for it to be successful.

The Trouble with Diagnosing Complicated Grief

A mental disorder shares a fuzzy boundary with normative human difficulties. We recognize a mental disorder when a person hears voices that make them believe terrible things about themselves. We recognize a disorder when a person's crippling anxiety prevents them from leaving their home. When a person cannot remember their loved one's names, or when they suffer so much psychic pain they wish they were dead, we can identify these states as mental disorders. Psychologists and researchers are working hard to understand and explain the murky boundary between disordered grieving and the universal human pain of loss, through enumerating specific diagnostic criteria, through assessment of functioning in daily life, through exclusions for how long ago the death happened, and whether the reaction appears conventional through the lens of the person's culture.

For people who are grieving, who have never before felt the rending pain of losing a loved one, using the term *complicated grief* may provide a way to convey how awful they are feeling. But suffering accompanies typical grieving, even when it is not a disorder. I worry that people will describe themselves with the term *complicated grief* because they believe that the depth of their grief cannot be normal, and the fact that the undertow of grief persists cannot be normal. But this is a common concern—grieving does take time, and restoring a meaningful life takes time, in the most normal and natural cases. I worry about overdiagnosis, from professionals and from grieving people themselves, who are simply trying to explain their experience in a culture that does not understand the universal grieving process.

I have seen the term *complicated grief* adopted by people like a badge of loyalty to the deceased loved one, a description of how deeply they loved. But connection to the universal nature of grief helps connect us to our fellow human beings, so a diagnosis needs to be used carefully, in cases where the complications uniquely require intervention. Having the term as a clinician allows me to telegraph to colleagues and insurance companies that this grieving person requires intervention to get back on the trajectory of healing. The diagnosis allows us to use the carefully honed and empirically studied psychotherapeutic treatments that create an on-ramp back to a meaningful life for those who suffer from complicated grief.

CHAPTER 6

Yearning for Your Loved One

The moment of separation from a loved one can feel like your heartstrings are being pulled from your chest until they will snap. These attachment bonds, these tethers, are both invisible and intensely real. They keep us connected to our loved ones; motivate us to return to them, like a pliable elastic band; and create a feeling that something is missing when we are apart.

My own vivid experience of such a separation moment from my spouse happened in my mid-twenties. I was newly married, only a few months after our wedding, and my mother was in hospice. My wife and I lived in Arizona, where we were in graduate school, and my mother lived in my childhood hometown in Montana. As is often the case in terminal illness, my mom had one medical crisis after another, and I flew to see her frequently. I had been fly-ing since I was eighteen months old—my mother was British and my whole maternal family lived in England, so my childhood was full of transatlantic flights. But because of the intense emotion sur-rounding the flights while my mom was so sick, and the upheaval from which I always seemed to be coming or going, I developed a terror about flying. When I got on an airplane, I felt full-blown panic. I did embarrassing things to get through landing and tur-bulence, like rocking in my seat and singing to myself under my breath.

In December 1999, my mother had a final medical crisis. My sister had already flown home, and it was recommended that I re-turn home, too. My wife and I decided that it made more sense for her to stay in Tucson, to wait and see if this hospitalization was just another in the string of events. She would follow me in a few days, if need be. As I boarded the last flight I would take while my mother was alive, forcing myself to leave the person to whom I felt closest in the world, and willing myself to enter the horror that was that plane—it was like ripping the sinews between us. Despite the fact that the decision was the right one, all the machinery of my brain was screaming at me not to leave her. Powerful chemicals and neural connections tried to prevent me from leaving the safety and love that I knew. Even with the good fortune of knowing I would see her again, I will never forget that powerful feeling of separation.

Aching for a loved one while they are alive but far away is use-ful for maintaining our bond with them; the ache can become un-bearable when we know they will never return. People describe the overwhelming pain of grief, beyond individual emotions, as psychic pain. Why does grief hurt so much? My studies of the brain have considered this question, and I believe that the brain has powerful tools, including hormones, neurochemicals, and genetics, to produce this aching and seemingly unbearable sensation.

Who Are You Again?

I want to take a little detour before we answer the question of why the loss of a loved one hurts so much, in order to tell you about how the brain identifies that particular loved one in the first place. In order to figure out which person we feel terrible about leaving, the brain faces an interesting problem. For most of us, in the humdrum routine of our lives, going home after work does not require much contemplation. However, it may surprise you to learn that the brain has to devote memory disk space to remember the exact same member of our species to partner up with night after night. It has to remember that this particular human being is the one it should go home with after dinner, and not that other good-looking one you have noticed. Your loved one does not look the same on the day you fall in love with them as they do a decade later, or the decade after that. Yet we feel pretty certain that this is the same person we met and married, or was born to us and we raised. In fact, an entire brain region, the fusiform gyrus, specializes in remembering human faces and identifying and remembering which person is *your* person. Neuroscientists have determined that this is the brain region where that thinking occurs, because people who experience a stroke or head trauma that affects the fusiform gyrus lose the ability to recognize familiar faces. This condition, prosopagnosia, prevents them from recognizing even someone as familiar as their husband or wife.

The idea that the fusiform brain area is dedicated to recognizing faces, or the *face-specificity* hypothesis, has seen a great deal of debate and investigation since the late 1990s. An alternative, the *expertise hypothesis*, originates in experiments done by psychologist Susan Carey and neurologist Rhea Diamond. The expertise hypothesis suggests that this brain area may specialize in recognizing any example of a category, such as a Mini Cooper or a '57 Chevy as examples of cars. You could imagine that in experts, like car aficionados or longtime dog show judges, this brain area could be especially tuned to particular categories. These experts would need to make fine discriminations among the larger categories of "cars" or "dogs." The expertise hypothesis suggests that although the fusiform gyrus is specifically recruited when looking at faces, this is because all human beings are experts on faces. Humans need to recognize specific people in many different situations, under different lighting conditions, and from different angles, just as expert dog show judges need to identify specific animals even within a species. Human face training, which makes us all experts, happens even early in infancy, when vision is best at the eight- to twelve-inch distance that brings our caregivers into focus as they cradle us in their arms. Our social world demands that we keep studying faces throughout development and adulthood. The debate between whether the fusiform gyrus functions to detect only faces, or specific examples of any category of objects, is not yet settled.

But, though the debate is not settled, there is good reason to think that this particular area of the brain is prepared from the beginning to learn faces. Some of that evidence comes from the fact that people with brain trauma in the fusiform gyrus—people with prosopagnosia, who are unable to identify faces—are still able to discriminate individual objects in other categories. On the other hand, people with brain trauma that does not affect the fusiform gyrus cannot expertly identify objects, but can still identify faces. For example, a patient identified as "CK" suffered a closed-head injury and was tested on his recognition ability.[1] CK had a collection of thousands of toy soldiers and complained that he could no longer distinguish an Assyrian from a Roman from a Greek soldier, much less identify specific soldiers within an army. Yet his human facial detection of friends and family was as good as anyone else's.

In our first neuroimaging study of grief described in chapter 4, the fusiform gyrus was activated when bereaved participants saw photographs of their loved one, compared to looking at photographs of a stranger. Presumably, we make a close investigation of the face of a loved one for whom we are grieving, and doing so relies on this brain area. It is significant that people did not use the fusiform gyrus area associated with facial recognition when they looked at the words that reminded them of their deceased loved one, also suggesting that the area is specific for faces, not other reminders of the person.

Single Brown Vole Seeks Mate

We have established that the brain can identify *who* our loved ones are, so the next question is, *Why* do we choose to return to them over and over again? And why does it hurt so much when we cannot find them? We actually know quite a bit about how the brain prompts seeking-my-mate behavior because of unique rodents called voles. Or rather, two different kinds of voles. Prairie voles live all over the plains of North America, whereas montane voles live at higher elevations in the western United States and Canada. What brought these two mammal species to the attention of scientists was that prairie voles are monogamous, while montane voles are polygamous, despite being very similar genetically. Although much has already been written in the popular press about the bonding of these furry little animals, scientific work since 2007 has also looked at what happens when voles are faced with permanent separation from their mate.

First, let's look at the prairie voles' mating habits. For a monogamous prairie vole, one day they meet another vole on the market, and after a head-over-heels day of mating, they are profoundly changed. Now they ignore other voles, prefer each other's company, build a nest together, and eventually play equal roles in parenting their pups. This is a pair bond for life. For voles that lifespan is about a year, although they can live up to three years in captivity. Neuroscientists Larry Young and Tom Insel (who later became the director of the National Institute of Mental Health) had a hunch that this permanent change after bonding was related to two hormones released in the brain, oxytocin and its close chemical cousin, vasopressin. In order to test whether these hormones were critical to the neural development of a bond, they blocked oxytocin during that initial day of mating. The prairie voles still mated, but they did not develop a preference for each other; in other words, they did not develop a pair bond. In a different test, the researchers put the prairie voles together but did not let them mate. If they gave them oxytocin (for the female) and vasopressin (for the male) during that time, the couples formed an enduring pair bond despite maintaining their virgin status.

Montane voles are much less social in general than prairie voles, and do not have mate preferences across time. When given these same hormones, the polygamous voles still did not develop a pair bond with each other. This is where the regions of the brain come in. Although both types of voles have receptors for these hormones, the receptors are in slightly different parts of the brain for prairie and montane voles. The monogamous prairie vole has more of its receptors for oxytocin in a part of the brain called the nucleus accumbens, compared to the montane vole. We will see later in this chapter that the nucleus accumbens region in the human brain is important for bonding in people as well.

Lock and Key

The hormones oxytocin and vasopressin play an important role in the neural mechanisms supporting pair bonds. These chemicals act like a key in the brain's lock-and-key mechanism, and the receptors for oxytocin and vasopressin are the lock, or keyhole. The number of receptors can vary for many reasons, differing between species, between individuals, and in response to events in one's life. Oxytocin could be flooding the brain, but if there are not enough oxytocin receptors for the oxytocin keys to fit into, the chemical flood will not have any impact on the neurons and the connections between neurons, and therefore, will not affect our thoughts, feelings, and behavior.

Chemicals and receptors are made by genes. Genes are the cookbook for how to make everything in the body. However, enzymes prevent some of the recipes from being made at any given time. These enzymes are involved in the epigenetic process ("epigenetic" means "near to the genes"). The enzymes are like the wrapper on a cookbook, holding it partly closed so that fewer gene recipes can be made. Under certain circumstances, this wrapper is removed. That certain set of circumstances, for prairie voles, is hanging out and mating with the newfound one-and-only for the first time. Having sex releases hormones, bathing the brain in oxytocin and vasopressin. The enzyme wrappers around the cookbook are removed, so more oxytocin receptors can be made, increasing the number of locks that the oxytocin keys can engage. This all has to happen while the vole is looking at, smelling, touching, and interacting with their new love, so that the new neural connections and associations are made for the sight, smell, and feel of this one very specific vole. (I am sure that the earth also moves and time stands still for the voles during sex, but that's harder to measure.)

Through some clever experiments, we know this is how bonding works.[2] Researchers put a drug in the nucleus accumbens of prairie voles while they were hanging out together for the first time, during one of the experiments where they did not allow them to have sex. This drug released the wrapper, so that the gene recipe could be "read" to make additional oxytocin receptors. The oxytocin receptors increased, just like when voles had sex on their first date, and the voles pair-bonded. The combination of the vole being present, and the vole brain being bathed in oxytocin, increasing their receptors, leads them to form pair bonds. The mate has to be present during that time so that the memory and knowledge of this particular vole becomes stamped in their brain, in their very epigenetics.

Once the wrapper has been taken off the cookbook, it usually remains off, and so the changes that support bonding endure. This is a permanent epigenetic change. Important experiences, like having sex for the first time with a partner, can change whether we *use* particular genes (following our metaphor, this would be the equivalent of making the recipes). If the wrapper remains on the cookbook, not as many oxytocin receptors are made, even though the gene was there all along. Mating can change other behaviors, like wanting to build a nest together on the Upper East Side and walking your vole children to school together. This permanent epigenetic change is what motivates us to return to this specific mate over and over again, recognizing them as our one-and-only. Once we are with them, the nucleus accumbens has other chemicals it deploys in the service of our bonds, including dopamine and opioids, that make us feel good together. Not only do we recognize them when we return, but it also feels good to come back to them again and again.

Meet Me in New York

In 2015, I was invited to attend a workshop at Columbia University in New York City. Neuroscientist Zoe Donaldson, now at University of Colorado, Boulder, brought together a small group of investigators working on the neurobiology of grief from different perspectives. Donaldson and a couple of the researchers studied voles, and a couple of us were clinical neuroscientists. We each presented our work, trying to translate our findings across disciplines. That night we ate sushi together in Manhattan, continuing our stimulating conversation. We wondered if we could measure grief in a rodent. Donaldson put it this way—how do you measure how an animal feels about the absence of something? That question has continued to propel our little group of neuroscientists to search for the important aspects of adapting to loss in animals and in humans, from the perspective of the brain.

One of the investigators I met in New York was Oliver Bosch, a neuroscientist from the University of Regensburg in Germany. He has done groundbreaking work, observing what happens to the pair-bonded vole when separated from their mate. More than that, his elegant studies provide more mechanistic detail about the brain systems that change when this happens.

As Bosch points out, for any social mammal from humans to chimps to voles, being in isolation is stressful. Over and above general social isolation, a particular stress response occurs when you separate animals, including humans, from close kin. Upon separation from their mate, voles make more of a hormone that is very similar to human cortisol, a stress hormone. The separated vole also makes more of the hormone in the brain that stimulates the release of rodent cortisol, the corticotropin-releasing hormone (CRH). This separation is all made worse by the fact that usually their mate would care for them when they got home at night after a stressful day. Ordinarily, following a stressful situation, when voles return to their nest, a male or female partner consoles them through licking and grooming. I have heard bereaved people describe this in their own way, saying that the extraordinary stress of grieving feels particularly awful because they are facing it without the one person they would usually turn to in difficult times.

I had the good fortune of visiting Bosch at the University of Regensburg, where he told me a fascinating extension of the vole story. What I find particularly interesting is that once the voles are pair-bonded, their brain system is primed, ready to make the CRH hormone if their mate goes missing. That way cortisol can quickly be released when they lose track of each other, motivating the vole to seek out its partner in order to reduce the resulting stress. Bosch described this as a gun being cocked when bonding happens, and separation then pulls the trigger. He told me that this increase in CRH in the rodent brain during separation also prevents the oxytocin locks and keys from working properly in the brain. Usually, when the little vole couple is reunited and oxytocin kicks in, the stress hormones return to normal. In bereavement, the physiological stress continues without the input of the pair-bonded mate.

Enduring Grief

Of course, with an additional two pounds of brain, humans have a significantly more complex system for bonding than voles. Similar primal mechanisms are probably working in the background in people, but they are considerably regulated and reshaped by our large, evolved neocortex. For most of us, when we are with our loved ones, we primarily feel safe and comfortable, rewarded by chemicals released in particular brain areas when we make contact with the specific mate we recognize.

Our need for the people we love, our attachment needs, are so basic that people are at increased risk of early death if they are socially isolated.[3] Most of us can learn over time to have our attachment needs met in a new or different way. This happens through strengthening the bonds we have with other living loved ones, by developing new relationships, and by transforming the bonds we have with the person who has died. These transformed, continuing bonds allow us to have access to them at least through the virtual world of our mind. The people that clinical psychologists really worry about, however, are the group that cannot seem to pick up the pieces of their lives after the loss, those who have complicated grief. In my scientific work, I wanted to understand whether these two groups, those with a resilient trajectory and those with complicated grief, responded differently to reminders of their loved one who had died, and what might be holding those with complicated grief back from engaging more fully in their lives.

In my second neuroimaging study of grief, UCLA social neuroscientists Matthew Lieberman, Naomi Eisenberger, and I used the same task of looking at photos and grief-related words as I had in the first study. When we looked at everyone who participated as a group, regardless of how they were adapting, we saw a general replication of the first study. Many of the same areas of the brain were activated in response to the photos and words about the deceased loved one, like the insula and the anterior cingulate cortex, which are buried deep in the middle of the brain. As I described before, often these two regions are activated together, when an experience is painful, both physically painful and emotionally painful. It is probably more accurate to say that they are activated because pangs of grief are very notable or salient, and that their salience activates these regions, but it is useful to think about pain in relationship to grief, and many people perceive and describe grieving as "painful."

Before we go on to the difference in neural activation between the complicated and more resilient bereaved groups in the study, I want to share a few more things neuroscience can tell us about pain. Remember that part of physical pain is sensation, and there is what we might call the suffering part of physical pain, the alarm bells that go off when we feel pain. These alarm bells are the brain's way of getting our attention: "Hey, this is important! Stop touching that! You are going to do serious tissue damage!" You can think of this as the "salience" of pain, and the insula and anterior cingulate are involved in sending those messages. Social interactions can also be painful, like being rejected by someone or being discriminated against. Although we now know that emotional pain is not encoded in exactly the same neurons as physical pain, the areas that encode the salience (the sense that this is important, this is bad, this is serious) of both physical and emotional pain are very close together, and enable both experiences to include suffering.

This One Is Not Like the Other

When we looked at all the participants together in this second neuroimaging of grief study, we saw that everyone who was bereaved showed activated brain regions related to the salience, or the alarm bells, of grief. We also looked at differences between brain activation in the typically adapting, resilient group compared to the complicated grief group. In order to attribute any group difference to grieving, we made sure the two groups were similar in other ways. The groups were, on average, the same age, and the same length of time had passed since the death. The people in the two groups were all women and all had lost a mother or a sister to breast cancer. Another similarity among the study participants was that their loved ones had not died suddenly, but following illness and treatment over many months.

I met some remarkable people in this neuroimaging study. I vividly remember a middle-aged woman who had lost her sister to breast cancer. The two sisters were hairstylists with adjacent stations at a salon. They lived near each other and even took their vacations together. Although the sister in my study was married and had children, her older sister was the person to whom she felt closest in the world. Her sister's death devastated her, and she felt lost without the daily interaction with this person who had been in her life every day since she was born. She had treasured their relationship and knew how lucky she was. There was no way to meet someone now, in the present or future, with whom she would share that history. No one could ever know every day of her life, as her sister had. Her life felt so diminished by the loss as to be meaningless. This woman was experiencing complicated grief.

One brain region distinguished the complicated grief and resilient groups; it was the nucleus accumbens,[4] the same brain region important in developing the monogamous pair bond in voles. The nucleus accumbens is part of a network well known for its role in other processes of reward (more on this below), including responding to pictures of chocolate among people who have cravings for it. The group with complicated grief showed greater activation in this region than the more resilient group. During an interview prior to the brain scan, we asked participants to rate, on a scale of 1 to 4, how much they had been yearning for their loved one lately. Across all participants in the study, the higher the level of yearning they indicated, the higher the level of nucleus accumbens activation. We found that the length of time since the death and the age of the participant were not related to the nucleus accumbens activation. Even the amount of positive emotion and negative emotion that the participants were experiencing was unrelated to the accumbens activation. Only the yearning—the feeling of craving or pining—was related to this neural readout of the nucleus accumbens.

It seemed very strange that the group who was not adjusting as well, the complicated grief group, would have more activation in the network responsible for reward. To be clear, reward as used by neuroscientists is not just something pleasurable. Reward is the encoding that means, yes, we want that, let's do that again, let's see them again. Several human neuroimaging studies have shown activation in the nucleus accumbens when participants looked at photos of their (living) romantic partner or pictures of their children. The hairdresser would have shown activation in the nucleus accumbens when looking at a photo of her sister while she was still alive. So why is this activation greater in the group with complicated grief? We interpret the reward activation in those who are experiencing complicated grief in response to reminders of a deceased loved one as occurring because they continue to yearn to see them again, as we do for living loved ones. It seems that those with more resilient grief may no longer be predicting this reward-ing outcome as possible.

I want to be very clear here, because craving implies addiction, and addiction is distinct from what I am suggesting happens in complicated grief. Other researchers have suggested that we may be "addicted" to our loved ones, and in my experience, this is a stig-matizing description for people who are suffering from a loss. It is also not quite accurate. Let's think of other human needs, like food and water. We would describe hunger and thirst as motivational states that cause us to seek out food and water, but we would never say that someone is addicted to water. We would say that they des-perately need water. Thirst is a normal motivation that the brain developed to fill this basic need. Attachment to our loved ones is also characterized by the normal motivational state of yearning. I am saying that yearning is very much like hunger or thirst.

A Critical Look Backward

There are trade-offs between the scientific need to have a group of very similar participants and the desire to be able to apply the results to the population as a whole. The participants in our second neuroimaging study on grief were all women, middle-aged, and primarily white. That is not what the majority of people who are grieving in the United States look like, let alone in the world. But the most significant critique of my own study is that the neuroim-aging scans took place on a single day in a whole trajectory of days for these grieving individuals. Interpreting the study relies on an inference of how one scan fits into the many days before, but we cannot know if that inference is correct without scans done several times across the trajectory of adaptation during grieving.

The inference works like this. We know from previous imaging studies that the nucleus accumbens is activated in response to living loved ones, like one's romantic partners or children. We imagine that this would have been true for the people in our study as well, in the time before we met them, when their loved one was alive. In our grief study, those who are adapting well had stopped having activation in the nucleus accumbens region, and those with complicated grief continued to show the nucleus accumbens responding to these photos. The inference lies in the words "stopped" and "continued." Continuing implies a period of time, but what we got is actually a snapshot of a single time point in different studies with different participants. The idea that activation in the nucleus accumbens changes across grieving is a logical inference, which fits with the data and theories that we currently understand about grieving, but it is not empirically proven.

Because our understanding of the neurobiology of grief is in its infancy, opportunities for speculation are many. In acute grief, the brain enables us to learn about our new circumstances, to make more accurate predictions about our world, albeit with painful emotional responses to reminders of the deceased person. Perhaps the brain can give us insights into the course of chronic grieving as well; perhaps there are natural variations in the neural systems that ordinarily support grieving adaptation. If the oxytocin system is involved, perhaps those with complicated grief have more oxytocin receptors, or their oxytocin receptors are concentrated in different brain regions. Maybe this creates very strong bonds with living loved ones, which is a good thing, but when circumstances of bereavement require us to adjust to life without the deceased, perhaps the same oxytocin-linked mechanisms make it very difficult to shift our focus to other people in our environment.

An interesting possibility is that genetic variations in the oxytocin receptor might put people at risk for developing complicated grief. Some hints at this possibility include the relationship between particular oxytocin genetic variations and adult separation anxiety, and several studies showing a link between these genetic variants and depression.[5] Much more research with many more people needs to be done in this area before any conclusions can be drawn, however.

A Magnificent System

The brain's ability to create and maintain bonds is magnificent. Certain hormones are released during specific activities like sex, or giving birth, or nursing. Because those hormones flood the brain, and receptors are there, neurons in particular brain regions make stronger neural connections and perform their specialized mental function better following these experiences. This is called permissiveness, because the hormones released during the event give the neurons "permission" to create thicker or more sprouted neurons, or to build more receptors. Oxytocin in the nucleus accumbens permits strengthened attachment bonds, motivating you to seek out this person and not to seek out others. Oxytocin in the amygdala permits better recognition of others and better control over anxiety.

Oxytocin in the hippocampus permits better long-lasting spatial memory, at least in mice, probably in order to allow mothers to keep track of their wandering children.[6] This person that you have fallen in love with, whether it is your partner or your baby, has opened new pathways in your brain. To make it clear, it is not just the hormones that are doing this. If hormones are dumped into the brain when you are in a room by yourself, this bonding will not (cannot) happen. It is only when these life-changing experiences happen to us *while interacting with the other person* that we fall in love—we deeply encode and remember the way they look, the way they smell, the way they feel, and provoke us to yearn to find them, over and over again.

This deep encoding of our loved ones in our brain is powerful. It has a powerful effect on our behavior, on our motivation, and on how we feel. Encoding someone means that yearning is the inevitable result of separation from them. Our brain is doing everything in its power to keep us united with the ones we love. These powerful tools include hormones, neural connections, and genetics, which may even sometimes override the painfully obvious knowledge that the loved one is no longer alive. The magnificence of the brain gave me great empathy for what bereaved people overcome in order to make a life when their loved one will not return. Their adaptation requires the support of their friends and family, the passage of time, and some considerable bravery to overcome what part of our brain may think is best for us. There are, fortunately, other parts of our brain that animals like voles do not have. We can use these parts to help us navigate overwhelming emotions during grieving, and that is where we turn our attention next.

CHAPTER 7

Having the Wisdom to Know the Difference

After I discovered how important yearning is from the perspective of the brain, I became more and more interested in figuring out exactly what yearning is. I set out to study it systematically, and to do that, I developed a self-report scale with a variety of questions to characterize different aspects of yearning. Like many people, I was curious as to whether yearning from the death of a loved one was the same as yearning after a romantic breakup or yearning during homesickness. So, psychologist Tamara Sussman and I named it the Yearning in Situations of Loss (YSL) scale, and we phrased the items so that they could be used in all three situations.[1] For example, one of the statements is, "I feel like things used to be so perfect before I lost _____." That wording appears on the version of the scale for people who are bereaved, with each person filling in the blank with their loved one's name. For a romantic breakup, the statement is phrased, "I feel like things used to be so perfect before _____ and I broke up." For homesickness, the equivalent question is, "I feel like things used to be so perfect when I lived in _____."

We learned a great deal from this endeavor. At least among young adults, the amount of depression they experienced contributed to their yearning, statistically speaking. But there was less of an association between yearning and depression than between yearning and grief, for the bereaved group. Similarly, there was less of an association between yearning and homesickness (for the group that had moved away from home), or between yearning and protesting a breakup (for the breakup group) than between yearning and depression. This reminded me that although there are shared features between depression and grief, they are not the same. For one thing, there is no specific person or thing that people with depression are preoccupied with, or yearn for. Depression is a more global experience, a hopeless and helpless feeling that attaches itself to everything that is happening and has ever happened and ever will happen.

After the yearning scale was published, Harvard psychologist Don Robinaugh assessed yearning with the YSL scale in a much larger clinical sample of treatment-seeking bereaved adults.[2] In his study as well, yearning was more closely associated with prolonged grief disorder than with depression. The level of yearning did not vary by gender, race, or cause of death, although those who lost a spouse or a child exhibited higher yearning than other types of kin losses. Yearning was somewhat lower when a longer time since the loss had passed, suggesting that even for those seeking therapy, yearning may subside somewhat over time. With specific descriptions of the nuances of how people were feeling, we now had a better understanding of what it means to yearn for our loved ones.

Then Suddenly, Out of Nowhere . . .

Robinaugh also pointed out that yearning refers to feelings and thoughts, and our felt experience is often a mixture of both. Given how painful yearning is, I wondered why it is so insistent, and why we continue to think about the deceased loved one so much. I want to tell you what scientists have learned about these yearning thoughts, and then we will come back to the feeling of yearning.

The thoughts we experience while yearning have a specific quality to them. Let me give you an example from my own experience. Late afternoon on a Sunday, I had finished grocery shopping, and I was looking in the fridge, considering what to make . . . and then suddenly I could see my dad in his kitchen, planning one of his famous dinner parties, with invitations to other widowers in town and the promise of roasted chicken and endless mashed potatoes. Another time, I picked up the phone and called him to tell him about . . . and then I realized I would not be able to have that conversation with him, and he could not give me his undivided, loving attention in the way he used to.

Again and again, our loved one who has died is suddenly there in our mind. We find ourselves partway into a thought, and then they come to us, which makes us yearn for them. Sometimes we don't even know what sets us off. In fact, our first awareness can be of the feeling of grief, without any clear idea of where it came from. Psychiatrist Mardi Horowitz called these *intrusive thoughts* and described their occurrence in a variety of stress-response syndromes, like after the death of a loved one or another traumatic event. He explained intrusive thoughts are both common and disruptive in the early weeks and months after the event. Part of what is so upsetting about them is that they feel involuntary. These intruders take over unannounced, stealing the moments when you are doing nothing in particular, when your mind is wandering. Although it is reassuring to know that intrusive thoughts are normal, and almost always decline over time, new empirical studies have challenged some of our assumptions about them.

Intrusive thoughts are memories of personal events and people that come to mind suddenly and spontaneously, without our intending to recall them. Remembering the loss reminds us of how much we miss them, which leads to feelings of distress or grief. But are intrusive thoughts more frequent than other kinds of thoughts, or do they just feel that way?

In the case of grieving for my dad, I can recall many moments in which I chose to bring him to mind. In the weeks and months after his death, I reached out often to talk to my sister and the wonderful family friends who helped us care for him. We would reminisce about things he said or did toward the end of his life. Once, when his bed was being wheeled from one hospital room to another, the nurse moving him could not see a small trash can in the hallway and bumped into it. My dad looked up, grinned, and said in his mischievous way, "Women drivers!" We must have told this story a hundred times in the first few months after he died. This memory of his constant good humor when encountering difficulty still brings a smile to my face and a twinge to my heart.

The fact that I frequently spent time thinking about memories like that after his death calls into question psychologists' beliefs about intrusive thoughts, because, as I said, in this case I chose to remember the event. Danish psychologist Dorthe Berntsen asked people who had experienced a recent stressful life event about their thoughts during daydreaming or mind wandering. She discovered they had voluntary memories, like the one of my dad being moved in the hospital bed, as frequently as they had involuntary memories, like the one of my dad cooking in his kitchen that came to mind spontaneously.[3] Although the involuntary memories are more upsetting, they are not actually more frequent than voluntary ones. Recalling memories of both types is more common after a stressful life event than when life is smooth sailing. The involuntary ones feel more frequent because they bother us more, probably because we are not prepared for the emotions they bring up. So, even though telling the story of my dad's humor brought up strong feelings, it was not as upsetting because I chose to bring it up, and therefore was prepared for the emotional impact.

The distinction between voluntary and involuntary memories gets us to a key difference between the brains of humans and the brains of animals, such as voles. Humans have two extra pounds of cerebral cortex, but more importantly, most of it is located in the frontal lobes between our forehead and our temples. The frontal part of the brain is uniquely developed in humans and has many functions, including helping us to regulate our emotions.

Remember that when a memory is retrieved, it is like baking

a cake with many different ingredients located in multiple brain regions. We are using areas of the brain such as the hippocampus and the nearby areas that store associations with a particular memory. The brain also accesses visual or auditory areas to add realism to our thoughts, giving us the impression of seeing or hearing what we are imagining. These brain areas are all used when we have both a voluntary or an involuntary memory. To look at differences between these two types of memories, Berntsen carefully compared them in people having an fMRI scan. The area that was uniquely used during voluntary, controlled retrieval, as opposed to involuntary memories, was in the outer part of the frontal lobes closest to our skull, the dorsolateral prefrontal cortex.[4]

The capacity to intentionally bring something to mind is a human skill. It requires what neuropsychologists call "executive functions," like a CEO organizing and directing the other parts of the brain to carry out tasks. In many ways, the brain is generating memories in the same way whether they are intentional or intrusive. The difference is that for intentional ones, our executive control in the frontal lobes gets involved to instruct us to remember them.

Remember for a moment your college graduation, or the birth of your first child, or your wedding day. You likely thought about these events spontaneously in the weeks, and months, and even years following, even when you were not intending to think about them. These wonderful memories probably sprang to mind even when you were doing something mundane, or when you saw something that reminded you of that day. Intrusive thoughts arise for extremely emotional events, including those that are positive— they are not reserved for extremely negative events. But because intrusive memories of negative events upset us, we worry about what these unwanted thoughts mean about our mental health. Most of the time, and especially in acute grief, intrusive thoughts are simply what the brain does naturally, in order to learn from these important, emotional events.

When considered from the brain's perspective, our brain is accessing our thoughts of our loss over and over again. It does the same for important positive events. It is still unpleasant to be caught off guard and have your thoughts and feelings turn to grief. But your brain is bringing them up in order to try to understand what happened, in the same way that you may share memories and stories with friends to talk them through and gain a deeper understanding. When you think of intrusive thoughts this way, it feels more normal that they happen: your brain is doing this for a reason. They seem more functional, and less like a sign you are not handling your grief well.

Remembering Not to Leave the Baby in the Car

Involuntary memories happen all the time. They happen more if you have recently experienced a trauma, but they can come anytime. In the normal course of events, your brain randomly intrudes with specific memories, or even conjecture about the future, without your intentional permission.

Today, how often did you think about your spouse or your kids? At random times, do your thoughts turn to the lunch money you meant to put in your daughter's backpack? Do you remember to text your wife to see how the meeting with her new boss went? Our brain is constantly generating reminders. It is an organ built to manufacture thoughts the way the pancreas manufactures insulin. These push notifications from our brain intrude on our consciousness whenever our mind wanders, and they help us to remember those things that are most important. This is how we remember, for example, not to leave the baby behind in the car seat when we do autopilot tasks like grocery shopping.

I speculate that just as reminders about our loved ones spontaneously arise during our lives together, reminders will also continue to intrude on our thoughts after they are gone for a period of time. During bereavement, however, these same reminders bring the realization that they are no longer with us, and these pangs of grief catch us off guard when they arise. As our mind wanders, we continue to get reminders from the brain to call or text our loved ones, but now these reminders conflict with reality. Seeing these intrusive thoughts from the perspective of the brain may make them less worrisome. You have always had intrusive thoughts about your spouse, your kids, or your best friend. The emotional impact of them is different now that they have died, but being reminded of our loved ones is the nature of having a relationship. You get reminders because these people are important to us. That does not change right away because the person has died. Your brain has to catch up. It is still running its regular programming of sending out notifications. You are not losing your mind; you are just in the middle of a learning curve.

You Have Options

Now let's get back to the *feeling* of yearning. Imagine you are a young widow, sitting at the breakfast table alone, drinking coffee at the beginning of the day after your kids are off to school, and you are missing all the mornings you sat there with your husband, mornings you will never have again. This is a classic example of yearning. At its most basic, yearning is wanting the person to be here again now. The brain is producing a mental representation, a thought, of the person who is absent. That thought produces a feeling of wanting, a desire for them to be here. The thought and the feeling together are the components of yearning, and together, these form a motivational state. Motivation, however, can lead us to do a variety of different things.

In response to her yearning, one possibility is that the young widow throws the coffee cup across the room, storms out, and vows never to sit at that table again. This would be a pretty dramatic example of avoidance. Avoidance can be behavioral, where we avoid situations or reminders of the loved one or of death, or it can be cognitive, where we attempt to suppress thoughts of the person or of our grief—or a combination of both. A different possibility is to engage even more deeply in the daydream about your husband: how he looked, how he would have laughed, how he held his coffee cup just so. This might feel soothing, imagining him there, gazing at you. You might hear what he would say to you now, sitting there, miserable in your grief. Would he come up behind the chair and put his arms around you? Would he tell you to get up and get moving, that the day will not wait for you?

A third possibility is that you return in your mind to the night he died, going over the details as you have so many times before, in excruciating detail. That night you took him to the hospital, because he had been complaining all evening that he had chest pain, and you suddenly realized he looked ashen and sweaty. Why did you not consider it might be a heart attack, why did you believe him when he said it was heartburn from dinner? Why didn't you insist on taking him sooner? Why did he keep smoking, even after his doctor told him it would increase his chances of heart disease? Why did you not confront him? He might never have died if only you had been more insistent, if you had acted sooner.

In the example of the daydream as a response to yearning, your brain is orchestrating an experiential simulation, a virtual reality of how things could be now, contrasted with how they actually are, sitting there alone. By generating the "what ifs" in response to yearning, your brain is imagining events that might have played out very differently than they actually did. The alternate reality your brain vividly dreams up, where he did not die but is here with you, is unfavorably contrasted with the present moment in real life. In acute grief, these "what if" responses to the pangs of grief are common and completely normal.

Of course, there are many other possible responses, like calling a friend on that lonely morning, or going for a run to take your mind off things. In fact, the dual process model clarifies that healthy grieving includes many different responses, appropriate in different situations, at different times, and to achieve different goals. If you have to get to work, maybe throwing your cup across the room in order to break out of your reverie and leave the house is not the worst thing in the world. That would be an example of oscillating from loss-oriented coping to everyday life experience. Calling a friend for support, and deepening a relationship with someone you trust and who cares about you, could represent an oscillation from loss-oriented coping to restoration-oriented coping. This would reflect the greater importance this friend plays in your life now and will in the future. Ruminating about the day of his death might be seen as an example of exploring loss-oriented coping, allowing the reality of what happened that day to sink ever more deeply into your knowledge banks. What is important is the benefit of having many ways of responding to yearning that fit the situation and forward your goals, both at that moment and in the longer picture of adaptation.

Flexibility

In a study of grieving people's facial expressions, scientists found that people show a wide range of emotions when talking about their relationship with their deceased loved ones. After videotaping interviews with bereaved participants, researchers analyzed their facial muscle movements, finding fear, sadness, disgust, contempt, and anger.[5] Positive emotions were also quite common: 60 percent expressed enjoyment at some point, which included the crinkling around the eyes that signifies a "true" smile, and 55 percent expressed amusement. These were fleeting facial muscle movements, so the bereaved person did not necessarily register experiencing all of these feelings in the five minutes they were being videotaped.

To prevent interpretations of the facial expressions based on the viewer's expectations, the person coding the facial movements did not know the participant was grieving.

The frequency and intensity of people's feelings typically increase after a loss, like turning up the volume dial. It is not uncommon to hear people who are grieving say that it is the worst they have ever felt, or that they did not know they could feel so bad. Such emotional intensity forces us to deal with these new experiences. Regulating one's emotions becomes a necessary part of daily life. Psychologists, friends, and family often have strong opinions about the best ways to cope. Confronting one's emotions and understanding them has been considered a good coping strategy. Suppressing one's feelings, and avoiding thoughts that bring up emotions, on the other hand, has been put in the category of bad coping. The most recent research suggests that the subject is not that straightforward, however.

The most reliable predictor of good mental health is having a large toolkit of strategies to deal with one's emotions and deploying the right strategy at the right time. It can be exhausting to have such high emotional intensity in the initial grieving period. There are good reasons to ignore our grief some of the time, in order to give the brain and the body a break, or even to give a break to those around us who feel emotional contagion. Distraction and denial have their usefulness. Rather than asking which are the best strategies, the more appropriate question might be whether using a particular strategy is counterproductive at a given moment or in a specific situation.

For those of us suffering from complicated grief, it can be more challenging to moderate the expression of our feelings than for those people who are adapting more resiliently. Moderation can mean amplifying or dampening our feelings. This means it can be more difficult for us to really focus on our feelings to better understand what's going on or to calm ourselves down. Ultimately, this calls us to be more flexible. When we don't deal with our feelings flexibly, we may start to feel numb or incapable of describing our truest feelings, and these modes hinder our ability to connect with those around us: if you are numb or cannot express your deep sadness, you are less likely to receive the support and comfort you need.

If we never allow the feelings of grief to surface, and we cannot contemplate them, or accept them, or share them, they might continue to plague us. Each individual is different, and there are no rules that every single person can use to adapt during grieving. But flexibility in our approach and openness to dealing with feelings as they arise give us the best opportunity to regulate our emotions in a way that allows us to live a vibrant and meaningful life.

The Bright Side of Life

Let's say you know four bereaved people. One of them chooses to go to a party with friends, and another one decides to stay home to watch a favorite movie. A third person spends some time with family telling stories about the loved one who died, and a fourth one writes in a journal about their grief. Which of these four people would you be most interested in meeting and which one do you think is most like you? How appropriate do you think each activity is, and how do you think the bereaved person would feel after engaging in it?

These questions were part of a study done by Melissa Soenke, a social psychologist at California State University, Channel Islands, and Jeff Greenberg, a social psychologist at the University of Arizona. If you liked the last two people better and thought the activities that they chose were most appropriate and effective, you are like the majority of people who participated in the study. The last two activities, which involve confronting negative emotions in response to the death of a loved one, are often called *grief work*. In the Western world, they are typically considered the most appropriate and most effective ways to cope. Ironically, engaging in activities that typically raise positive emotions, such as going to a party or watching some form of entertainment, are actually more effective at reducing sadness and grief.

The "undoing" of negative emotions with positive emotions works because positive emotions change cognitive and physiological states. Positive emotions broaden people's attention, encourage creative thinking, and expand people's coping toolkit. Psychologists Barbara Frederickson and Eric Garland describe this as the upward spiral triggered by positive feelings. In a second part of the study by Soenke and Greenberg, bereaved participants wrote about their loss and then watched a funny clip from a television sitcom, worked on a word search puzzle, or watched a sad scene from a popular movie. After participants finished completing the activity, they rated their current happy, sad, and guilt-related emotions. These ratings were compared with their ratings at the beginning of the experiment. In line with data from Frederickson and others, watching the funny clip decreased feelings of sadness associated with recalling a sad event, while neutral and sad activities did not. Although engaging in activities that usually lift our mood is effective, bereaved people are often reluctant to engage in them.

There are at least two reasons why we usually don't choose mood-boosting activities when grieving. First, doing fun things is not considered the "right" way to act, so we worry what other people will think about our choice. Second, we anticipate that doing something enjoyable after a sad experience will make us feel guilty. When we violate social norms or expectations, guilt is a common response. However, even though people anticipated that they would feel guilty doing something fun, no one in the study felt guilty after watching the funny clip. But the anticipation of guilt can deter people from engaging in enjoyable activities. Other research supports this finding that humans are pretty bad forecasters of how they will feel in future situations.[6]

I am not suggesting that when we lose a loved one, we should go to party after party in order to feel happy instead of sad. Flexibility, as I mentioned before, is beneficial, like contemplating what happened, feeling the gravity of our situation, expressing our anger or sadness, attempting to understand how our life story has changed, and more. But now we know that mood-boosting activities are beneficial in their own right, so we might allow ourselves to do something fun, and even encourage our bereaved friends and loved ones to do so. In any case, it's another option for our toolkit.

Caring for the Bereaved

If you care for someone who is grieving, emotional flexibility matters for you, too. The challenge for those of us who love a bereaved person is to accept the reality that someone we care about is hurting. The challenge for the grieving person is to accept the reality their loved one has died. It is heart-wrenching to watch, but grief is a part of life. This is a time when your dear friend or spouse or sibling must face the painful reality of mortality. By analogy, if we see a child who has fallen down because he scraped his knee, we run over, scoop him up, and kiss him, reassuring him his knee will heal, because we know the painful sting will eventually pass. Or we look over and smile at him, acknowledging he took quite a fall, and encourage him to get up and keep playing. Having compassion for those around us experiencing grief may also include comforting or encouraging them, flexibly responding to the moment.

If you are listening to your grieving friend and supporting them with the goal of taking away their grief, you will only be frustrated if they continue to grieve despite your loving care. Of course, there is a difference in having compassion for an event that is brief and over relatively quickly like scraping a knee, and for grieving that takes many weeks, months, and even years. It is still vital to provide support, love, and care, but not because it will take away their pain. It is vital because by witnessing, sharing, and listening to their pain, they feel love and we feel loving. In any given moment, though, we may still have to decide if it is wiser to hold them while they cry or to encourage them to get up and keep playing, because flexible approaches to strong feelings are the most useful.

It is our challenge, as a friend to those in grief, to keep offering love, while also finding support for yourself in your wider community. This is important because caring for someone who is in pain is stressful, in a variety of ways. You may feel guilty that you are not overwhelmed with grief and wonder why this terrible thing is happening to them and not you. Or you may also be grieving, and your bereaved loved one may not be able to support you right now. It might feel unfair that they should be getting all the attention, and we want to say, "But I have grief, too!" more than we want to provide loving-kindness to them in that moment. With patience, we can separate giving a grieving friend what they need in terms of attention and love, while also asking for what we need in order to soothe our own hurts.

Serenity Prayer

Yearning, anger, disbelief, and depressive moods decrease across time after the death of a loved one.[7] These feelings do not follow stages, and people still experience them years after their loss. But their frequency declines as the frequency of acceptance increases. Acceptance may be the outcome of learning that a new reality is here to stay and that we can cope with it.

What we spend time thinking about matters. How we react to what we are thinking about, and what we feel, matters. How we handle what our minds do moment to moment can help. These insights remind me of the Serenity Prayer. Inherent in that plea for help is a recognition we have to flexibly deal with the trials we face: *God, grant me the serenity to accept the things I cannot change,*

*courage to change the things I can, and wisdom to know the differ-
ence.*

We cannot change mortality. We cannot change the suffering
that accompanies loss. We cannot change intrusive thoughts and
waves of grief. But if we have great courage, we may be able to
learn to respond to these indisputable circumstances with greater
skill and deeper understanding. The challenge is, of course, the
wisdom to know the difference, learning when to pause and reflect
and when to push on. The mysterious and overwhelming feelings
of grief require wisdom, but wisdom is gained through experience.
We turn to our loved ones for what wisdom they can give us. We
may turn to our spiritual or moral values to guide us. Finally, we
wait for our own brain to develop the wisdom to discern the best
course of action that comes with learning from each new day's experiences.

PART TWO

The Restoration of
Past, Present, and Future

CHAPTER 8

Spending Time in the Past

I n the 1993 movie *Fearless*, Jeff Bridges and Rosie Perez portray strangers who survive an airplane crash. Both of their lives unravel as they grapple with what it means to have survived. One night as they sit in his car, Perez reveals that she believes she killed her infant son by letting go of him during the crash. Bridges responds initially with utter frustration. When she completely comes apart, sobbing and praying to the Virgin Mary for forgiveness, Bridges is overwhelmed with what it must be like to believe what she believes, to feel like the murderer of the child given to you to protect. He gets out of the car, and with unclear intentions, tells Perez to get in the backseat and buckles her seatbelt. From the trunk, he retrieves an oblong, rusty toolbox and puts it in Perez's arms, telling her to hold on to it and that it is her baby. In what might be a suicide attempt, Bridges gets behind the wheel and drives them through an empty alley toward a concrete wall, with the speedometer ticking upward. He tells Perez this is her chance to hold on tight, to save her baby. Completely immersed in the scene that resembles the plane crash, she kisses the toolbox. The speeding car crashes into the wall, and the rusty orange toolbox flies like a rocket through the front windshield of the car and into the cinderblock wall, its steel metal crumpling. For Perez, it is immediately and palpably clear that there was no way she could ever have held on to her baby, no way she could have saved him. Through this immersion she realizes what actually happened, and the difference between the reality and her belief about what happened.

Psychologists call our thoughts about what could have happened *counterfactual thinking*. Counterfactual thinking often involves our real or imagined role in contributing to the death or the suffering of our loved one. It is the million "what ifs" that roll through our mind: *If I had done this, he never would have died. If I had not done that, he never would have died. If the doctor had done this, if the train had not been late, if he had not had that last drink* . . . The number of possible counterfactuals is infinite. Their infinite nature gives us endless thoughts to focus on, to consider and reconsider, turning the scene around and around in our mind.

The irony is that this type of thinking, creating the myriad situations that could have happened, is both illogical and unhelpful in adapting to what has actually happened. Our brain may still be doing it for a reason, however. Some would say the reason is to try to figure out how to avoid deaths in the future, but it may be simpler than that. Our brain, by focusing constantly on the limitless number of alternatives to reality, is numbed or distracted from the actual, painful reality that the person is never coming back. Even when the counterfactual thinking involves the painful experience of guilt or shame, like believing we killed our baby, our brain still seems to prefer it over the terrifying, gut-wrenching truth that our loved one is no longer here. Or, mulling over these counterfactuals can become a habit, a knee-jerk way of responding to pangs of grief. Although we are trading painful guilt for equally painful grief, at least guilt means we had some control over the situation. Believing we had control, even though we failed to use it, means the world is not completely unpredictable. It feels better to have bad outcomes in a predictable world in which we failed, than to have bad outcomes for no discernible reason.

The illogical nature of counterfactual thinking can be demonstrated like a geometric proof. Human beings make a common error in "if . . . then" statements. The "if" portion is called the antecedent; the "then" portion is called the consequent. Logicians use tree diagrams, like the one following, to figure out where the error of logic is. In the example of the young widow from chapter 7, she knows it is true that her husband died, and knows they went to the hospital in the middle of the night. She is subconsciously tempted to believe that because one antecedent (went to the hospital) is associated with one outcome (he died), the other antecedent (went to the hospital earlier) must be associated with the other outcome (he would not die). But that tempting logic does not make it true. It is not necessarily the case that if they had gotten to the hospital earlier in the evening, he would not have died. Of course, it is one possibility, but it is also possible that he might have died despite getting there earlier. We can endlessly consider what *could* be true in the counterfactual world where we wish we lived.

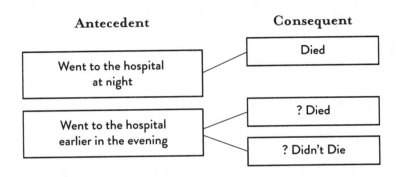

Some might think only an android like Data on *Star Trek* would think about the loss of a loved one this way. I was once talking about counterfactual thinking with a clinician who has worked with many people who have prolonged grief disorder. He agreed that it can be helpful to challenge a client's beliefs that lead them to feel extreme guilt. He also said he has been surprised, however, that revisiting the death during exposure therapy, in the context of a therapeutic relationship and without challenging the counterfactual thinking, often allows the "if only" thinking to just fade away. No need to explain the logic. Developing the ability to tolerate the strong feelings of grief, of helplessness, or existential loneliness brought up by the memory of the death, or by the realization that the loved one is truly gone, made the constant "what ifs" unnecessary.

Rumination

For some of us, a wandering mind lapses into worrying or ruminating. In worrying and ruminating, we are also imagining an alternative reality, in a similar way to creating "what ifs" during counterfactual thinking. Rumination focuses on things that have happened in the past, like ruminating over something we did wrong, or about how someone treated us. Worry focuses on events in the future, our anxious thoughts about worst-case scenarios. The process of these thoughts tends to be repetitive, passive, and negative. Psychologist Susan Nolen-Hoeksema defined the term *rumination* as a way of coping with feeling down by narrowing your attention to your negative feelings in an attempt to understand them. Nolen-Hoeksema was able to predict who was depressed, or would develop depression, by identifying people who spent more time ruminating.

In the last chapter I said that recalling memories of the loss and understanding our feelings of grief was helpful, and now I seem to be contradicting that by saying these thoughts cause depression. Well, the truth is, psychologists do not yet have all the answers to when (or how much) processing of thoughts about grief is helpful and when it is not. Researchers are actively grappling with the paradox that you cannot learn about what has happened, and therefore why you feel terrible grief, without focusing on yourself, on your sad and angry feelings. You cannot fully understand what has happened without letting your mind wander through the territory of rumination. At the same time, these ruminative thoughts can develop a life of their own, and when grieving people persist in these repetitive thoughts, they tend to develop complicated grief or depression. Although we do not have all the answers yet, some paths through the paradox are becoming clearer.

Rumination can be divided into two aspects, which Nolen-Hoeksema called *reflection* and *brooding*. An example of reflection is writing down what you are thinking, perhaps several days in a row, and analyzing your thoughts. Reflection is an intentional turning inward, engaging in problem solving in order to alleviate your feelings. On the other hand, brooding reflects a passive state. Brooding is finding yourself thinking about your mood even though you did not set out to think about it, and persisting in these thoughts even when you try to stop thinking about it. Brooding is passively wondering why you feel down or comparing your current situation with how you think things should be.

Nolen-Hoeksema studied the relationship between depression and both brooding and reflection, asking people to report on their thinking style and symptoms of depression.[1] People in this study were interviewed twice, about a year apart. The reflection aspect of rumination was correlated with having depression at the time of the interview. But reflection at the first time point was associated with less depression at the second time point. Brooding, on the other hand, was associated with more depression both concurrently and at the later time point. Notably, women tend to ruminate more than men, and women also have higher levels of depression. Women scored higher on both reflecting and on brooding than men, suggesting they are more contemplative overall. Only brooding, however, was associated with greater levels of depression in women. So, brooding is one link between gender and depression.

I think of this subtle distinction between brooding and reflecting as an emphasis on whether a person is seeking or solving. Seeking an answer might precede solving a problem, but feeling better usually requires getting to the solving part. Often, we feel better by settling on a solution to try, even if the planned solution does not ultimately fix things entirely. Feeling better requires stopping the seeking, or ruminating, or worrying, at some point. However, sometimes even problem solving can draw you back into a cycle of repetitive thinking and prolong your sad or anxious mood, unless you have the powerful capacity to continuously monitor your thoughts and change course as needed. This sounds like a task for a Zen master! We are able to strengthen the skill of directing our attention to our thoughts, however, and choosing whether our thoughts are helpful or not. This skill is often the focus of cognitive-behavioral therapy (CBT). But it does not come easily to most of us, especially after a death, when the powerful emotions of grief are prevalent.

Grief-Related Rumination

After my mom died, I ruminated a lot. In truth, I ruminated before my mom died as well, but after her death, feeling grief gave me lots of opportunities to focus on my mood. My thoughts would spin on why I felt down. I wondered if I was prone to depression because she had been. Or whether I would have turned out differently if she had not been depressed while I was a child. She leaned on me to help her manage her mood, and I was always afraid I would not be able to help her feel better. I learned that I was most successful at helping her feel better, at least momentarily, when I would say whatever she needed to hear or do what she wanted me to. This often meant I had to ignore what I thought or needed. The pattern of believing I should help her feel better at any cost became a well-worn groove. After her death, I repeated this pattern: I struggled to help other people in my life feel better, while I continued to ignore my own feelings. There were endless possibilities for why I felt down, and I carefully examined each one, prolonging the state I was in. It probably didn't help that I was in a clinical psychology graduate program, where I was training to examine people's moods and the causes for their feelings. Fortunately, I also learned many problem-solving methods and skills for improving mood, so I did not succumb to rumination all the time.

The mind ruminates when it cannot resolve the discrepancy between its current state, such as feeling down, and its desired state, such as feeling happy or content. During grief, the source of your awful mood is less ambiguous. When you feel the powerful yearning that accompanies grief for many people, the cause seems obvious. A loved one just died, and grief-related rumination focuses specifically on the causes and consequences of the death. In contrast, during depression, as I had after my mom died, rumination can be about anything. For people in acute grief, grief-related rumination focuses specifically on the death of the loved one, or the effect the death has had on a person. The death of a loved one intrudes on our thoughts, as we have seen, and the tendency toward rumination extends the length of time our thoughts stay on that topic. Rumination predicts depression, and grief-related rumination predicts complicated grief. People who had depression prior to a death often continue to be depressed afterward, as we saw in Bonanno's trajectories of grieving. Other people may not have been ruminators or depressed before, but the death may start the repetitive thought process. Psychologists now think that being unable to stop these grief-related ruminations may be one of the complications that get in the way of typical adaptation during grieving.

Grief-related ruminations tend to center on a few topics, as evidenced by Stroebe and Schut, and their colleagues, Dutch psychologists Paul Boelen and Maarten Eisma.[2] The five topics include: (1) one's negative emotional reactions to the loss (*reactions*), (2) the unfairness of the death (*injustice*), (3) the meaning and consequences of the loss (*meaning*), (4) the reactions of others to one's grief (*relationships*), and (5) counterfactual thoughts about the events leading up to the death (*what-ifs*).

Let's look at some examples. Often people worry about their own *reactions* to the death of a loved one, trying to understand the range and intensity of their feelings and whether those reactions are normal. Thoughts about the *injustice* of the death include feeling that the person should not have died and wondering why this happened to you and not someone else. Focusing on the *meaning* of the death includes thoughts about what the consequences of the death are for you, or how your life has changed since the loss. *Relationships* with friends and family are often affected by grief and loss, and these ruminations are about whether they are providing the right support, or the support that you desire. *What-ifs* are the counterfactual thoughts covered at the beginning of this chapter.

Studies of British, Dutch, and Chinese bereaved people show they all report ruminating about these topics. The more frequently they ruminate about these topics, the more intense their grief symptoms are. Not all the topics are equally problematic, however. In research on grief-related rumination, the first topic (ruminating about one's negative emotional reactions to the loss, or *reaction*) led to less grief at the moment and over time, at least in one study. On the other hand, ruminating on how others are reacting to one's grief (*relationships*) and on *injustice* were both associated with more grief at the moment and predicted more grief six months later.[3]

All of these topics of rumination are actually questions that cannot be answered, which is why they can persist indefinitely. There is no answer to whether the death was unfair, because there are many facets of unfairness. There is no answer to all the ways that their death has robbed your life of its meaning or joy, because losing a loved one brings an infinite number of changes. The sneaky problem with rumination is that while one is ruminating, it feels as though you are seeking out the truth of the matter. The point is that the thoughts are prolonging our sad or irritable mood, not whether the thoughts are true.

Imagine a family, stricken by the tragic death of a son by suicide. Nora feels devastated by the loss of her brother. On top of her grief, she feels even worse because there is a mismatch between her family's behavior and what she needs. She wants her family to acknowledge the pain he was in that brought him to the desperation of his decision. She wants them to acknowledge how this grief is especially painful for her, closest in age, inseparable in childhood. Her mother refuses to talk about him, and her cousins seem awkward and uncomfortable around her. Whether or not the family's reaction should be more open, more accepting and understanding of Nora's grief, is not the point. The point is that Nora feels trapped in an endless stream of thoughts, irresolvable and of no benefit to her. Ruminating, in and of itself, will not improve the situation. She may need to go into problem-solving mode instead, like talking to her cousins about what she would find helpful during this difficult time, or spending less time with her mother and finding friends she can talk with more openly. The trick is not to determine whether the thoughts are true, but rather whether they are helpful.

Why Do We Ruminate?

If we are ruminating in order to figure out what has happened and why we feel so awful, and yet ruminating does not actually help us adapt in the long run, why on earth would we do it? The answer may lie in what we are *not* doing while we are engaging all our cognitive resources in ruminating. Sometimes the subconscious motivation for engaging in an activity is that it allows us to avoid whatever else we might do, often because it feels better. To investigate the motivation for rumination, we might ask, how would we feel if we were not ruminating? Are we engaging in rumination because it feels better than what we would be doing otherwise?

Most of us do not like the experience of being overcome with grief. We feel slightly out of control; we may believe that if we allow ourselves to break down, we will never put the pieces back together again. It is painful, gut-wrenching. Stroebe and her colleagues formulated a remarkable hypothesis: letting our thoughts run through our mind again and again may be a way to distract ourselves from the painful feelings of grief. Thinking about the loss and the consequences of the loss might actually be a way to avoid *feeling* the loss. She and her colleagues called this the *rumination as avoidance* hypothesis.[4] This may sound pretty far-fetched initially, but luckily this careful research team did empirical studies to investigate. Let me tell you how.

When something is very difficult to measure, scientists develop special techniques to measure it—that was the basis for the microscope and the telescope. Avoidance is a difficult thing to measure. Although we can ask people how much time they spend ruminating, or what they ruminate about, it does not make sense to ask people outright about avoidance. If the brain's motivation for avoidance is not to notice what one is feeling, then avoidance itself as a process probably would not be noticed either. Special measurement techniques in the laboratory, however, allow psychologists to study automatic responses, responses too quick to be deliberate. These decisions are made by the brain very quickly. One method uses reaction time and the other uses eye tracking—responses that happen approximately as quickly as one heartbeat.

In order to test the rumination as avoidance hypothesis, Stroebe and her colleagues invited bereaved people to come to the laboratory and participate in these avoidance measurements. They thought using the picture and word composites from our neuroimaging study would work for them as well. This group of Dutch psychologists, Eisma, Stroebe, and Schut, contacted me and I explained how to create the composites that form four categories: photos of the deceased and photos of a stranger, each combined with grief-related words or neutral words. To measure reaction time, they asked bereaved participants to push or pull a joystick that made the photo/word increase or decrease in size on the screen, making it look like the photo was moving away from them or toward them. Tiny differences in the amount of time they took to push or pull can be measured in milliseconds. Our brain's automatic avoidance causes us to push a picture away a few milliseconds faster than it takes to pull it toward us. In addition to this laboratory task, participants in the study also reported on how often they ruminated on grief-related topics. The researchers found that bereaved people who ruminated more pushed the deceased/grief word picture away from themselves faster than bereaved people who ruminated less, and faster than they did for the stranger or neutral word categories.[5] These results suggest that more time spent ruminating is associated with stronger automatic grief avoidance.

In a different task, the same bereaved participants looked at the pictures on a screen while an eye-tracker measured the minuscule movements of their eyes, to determine where they were looking. The eyes are literally an extension of the neurons of the brain, a window into where the brain's attention is focused. In this study, two pictures appeared side by side. Those who reported ruminating more spent less time looking at the deceased/grief word picture than at the picture on the other side of the screen.[6] The ingenuity of these studies is that scientists would not be able to accurately figure out where a person focuses their visual attention just by asking the person. But the data clearly show that high levels of rumination are associated with the brain avoiding reminders of loss, either by pushing or looking away. Even though people ruminate about other aspects of the cause and consequences of their loss, they avoid these bold, frank composites that remind them of the fact of their loved one's death.

Perhaps you have experienced rumination as avoidance, without recognizing it as such. Have you ever had a friend who told you the story of her loss exactly the same way every time? She rattles off what happened, and tells you how awful it was. But you may feel like there is a disconnection between her telling you that it was awful, and the fact that she does not seem to feel awful in the moment that she is telling you. She may go on in great detail, and that level of detail is the rumination process, a cognitive process. Sometimes, telling the story in this cerebral, ruminative way allows us to avoid feeling what happened when our loved one died—rumination as avoidance. The trouble is that telling the story again and again in this way is not the same as discovering what the loss means. Discovering what the loss of this person means to us, on the other hand, and learning to find a way to live without them, would create strong feelings in us but also help us to grieve and to fit that loss into our ongoing lives.

So, rumination is an avoidance process, although not intentional. Repeatedly returning to aspects of the loss or one's grief that cannot be changed does not help us learn to tolerate the painful reality over the long term. I have known people who have told me that when they stopped trying to avoid feeling grief, grief was not as hard to tolerate as the effort required to avoid it.

As is often the case in our current scientific understanding of how the brain functions, we do not yet know if people who ruminate more do so because of the weaker connections between brain regions, or if ruminating leads to weaker network connections. As we often find in psychology, the answer is probably a combination of both, a downward spiral of form and function. However, a downward spiral often gives us the leverage to intervene and create an upward spiral. This upward spiral can be the skill learned in psychotherapy to attend to the content of one's thoughts and move one's attention to external features of our environment or do something to shake us out of our ruminative mood. For example, the young widow who threw her coffee cup and left the room managed to stop returning to her perseverative thoughts; she found an effective way to change what she was thinking about by getting out of the house.

In It Together

My best friend has been there for me through all the important events in my life, helping me face the death of each parent. She and I have written countless letters to each other over the years. Since high school, other than brief periods, she and I have never lived in the same place. Separation necessitated many letters, and eventually emails, and finally, with less time on our hands, phone calls. When I studied abroad in England, these letters got even longer and more important to me. I was terribly depressed during that year of college, and letters were an opportunity to reveal everything I was thinking and feeling. We enabled each other to articulate the nuances of our worst moments. I knew she understood what I was saying, and she could speculate on how my life and upbringing led to my feelings as clearly as I could. I genuinely do not know what I would have done without her.

It never occurred to me that this type of conversation could have both an upside and a downside until I read work by psychologist Amanda Rose at University of Missouri. She studies the role of these conversations, particularly in the lives of girls and young women.

She developed the term *co-rumination* to describe the repetitive, extensive discussion of personal problems between two close friends, an intimate and intense form of disclosure, often about negative feelings. The clear upside I experienced with my best friend was borne out in Rose's research. Friends experienced these conversations as increasing their feelings of closeness and satisfaction with their friendship.[7] On the other hand, co-rumination also led to increased symptoms of depression and anxiety. Support that involves talking about problems extensively may have a negative effect on emotional adjustment rather than a positive one. Ironically, this is a vicious cycle. When one feels more depressed, one may turn to these conversations more and more in order to feel close and supported.

The research is not saying that close friendships or disclosing one's feelings are bad. In fact, when Rose separated out the amount of co-rumination, these friendships are still associated with less depression. The opportunity to reveal one's inner life and find support and encouragement from another person is beneficial. The devil is in the details; passive discussion of the same negative feelings again and again is different from problem solving, encouragement, or advice. Talking about how you are feeling can make you feel normal when the other person has felt that way, too. When negative feelings are by far the most common topic you talk about, however, or when it feels that the whole world is against the two of you, it begins to slide into co-rumination. Over time, my best friend and I came to this same conclusion, on some intuitive level. She suggested we only discuss a particular situation three times, and if nothing had changed by then, we would try something new before we discussed it again.

Accepting

As I was writing this book, I had the great fortune to spend my sabbatical year at the University of Utrecht in the Netherlands. Utrecht is an old Roman city, full of people cycling along the many canals lined with beautiful flowers. I spent time at the historic university with my generous hosts, Stroebe and Schut. Working side by side with other grief researchers was a novel experience for me, as not a lot of scientists are devoted almost exclusively to this topic. In addition, living in another country provided an opportunity to absorb a great deal of art, history, and culture. Utrecht is famous for its Protestant history and theological pursuits. One day as I was thinking about the Protestant work ethic, I was struck by the word *work* in "grief work." Stroebe and Schut had been trying to puzzle out the differences between unhelpful rumination and useful grief work. It dawned on me that maybe there was an opposite of both rumination and grief work, and that might be *accepting*. I use *accepting* for one's response to what happens in the moment, rather than *acceptance*, which suggests a permanent change in how a situation is viewed.

As I imagined scenarios of confronting versus accepting a loss, one notable difference that struck me was the amount of effort required. Not that accepting is necessarily easy. But when accepting comes, it brings a certain sort of peace with it. It is like setting down something heavy, even with the full knowledge that you may have to pick it up again. And, although accepting may mean you are not consumed any longer by the thoughts and feelings surrounding the loss, accepting also feels different from avoidance. Avoidance— trying to circumvent the knowledge that the death has happened—

is effortful. Avoiding the overwhelming feelings of grief, motivated by how much you hate those feelings, requires effort. Accepting, on the other hand, does not have any bearing on whether or not you hate the fact that your loved one died. It simply acknowledges the reality, and stops the reaction there. No ruminating, no problem solving, no anger, no protest—just accepting the way it is.

To be clear, there is a distinction between accepting someone's death and resigning oneself to their death. Accepting is knowing that the person is gone, that they will never return, that there is nothing to be done about things that happened in their lifetime, that regrets and goodbyes are part of the past. Accepting is focusing on life as it is now without the deceased, without forgetting the deceased. Resignation goes one step further and suggests that your loved one is gone, and that you will never be happy again. It implies that there are only negative consequences to the death. Accepting is simple awareness of the reality, with the hope that the reality of the present moment can be meaningful or hard, joyful or challenging. Hope is a fundamental part of human psychology, when people are given enough support and time.

A few days after the death of my father, I went to Germany for about three weeks on a work trip that had been planned long before I had known he was going to die that summer. Fortunately, I was also working and staying with Gündel, my colleague and dear friend of twenty years, since our first fMRI study of grief. He is a psychiatrist and psychoanalyst by training, no stranger to grief and people who are grieving. Frequently, in the afternoons on that trip, I would find myself needing to cry. That was how I experienced it—one minute I was clacking away on my laptop and the next minute the floodgates would open, with tears welling up in my eyes. Losing my remaining parent was qualitatively different from losing one parent, in a way that I had not expected. It now meant that I had *no* parents; parents no longer existed in the world for me. I am not sure if the word *orphan* can be applied to a woman in her forties, but I felt very, very alone.

During these floodgate moments, I would get up and go for a walk, letting out my tears where I was not going to disturb my officemate or other people in the department. Southern Germany is beautiful in the summer, and that year was no exception. A walking path wound through a patch of leafy green trees behind the clinic, and I would walk there for maybe twenty minutes. This happened day after day, around the same time. I came to think of these crying spells like the afternoon rain showers that occur in some climates in the summertime. The sun is warm and smiling, and then all of a sudden there is a shower. Soon the sun comes out again, adding sparkles to the leaves and cars that are now wet. These summer storms are fairly predictable: not every day, but often enough that you remember to bring your umbrella or to look at the horizon before you head out in your sandals. There is no point in cursing these showers, no point in being upset when the rain falls in the middle of your perfectly good softball game or picnic. They are just going to happen, and they do not particularly care what you are up to at the time. I came to think of these afternoon crying spells in the same way: a familiar feeling when the dark clouds came over me, a somewhat predictable pattern in the afternoon, and the knowledge that they were unlikely to last. I would find myself at the end of the leafy circular path back at the clinic, and usually find also that I had stopped crying. My brain was back to thinking about some paragraph I had been writing in the office, or making a grocery list for dinner.

The key to accepting is not doing anything with what you are experiencing; not asking what your feelings mean, or how long they will last. Accepting is not about pushing them away and saying that you cannot bear it. It is not about believing that you are now a broken person, since no one can bring your parents back and you will never get another set. It is about noticing how it feels at that moment, letting your tears come, and then letting them go. Knowing that the moment of grief will overwhelm you, feeling its familiar knot in your throat, and knowing that it will recede. Like the rain.

Realization

Through understanding scientific research on mind wandering, by having people tell us what topics they ruminate about, and measuring their thought processes with reaction time and eye-tracking laboratory tasks, I realized that restoring a meaningful life requires flexibly moving our attention from thinking about the past to thinking about the present and the future. It requires being able to move our thoughts from relationships that were, to relationships that are and relationships that could be, and back again. We may still spend time in reveries about our life together, and a grieving trajectory certainly does not mean that we forget our loved ones who have died. In fact, the time we spent together, and the experiences we had bonding with them, resulted in neural connections and chemical consequences in our brain that will never allow us to forget. Choosing to spend time thinking of someone you care for now does not mean forgetting someone you loved intensely, and whom you will love forever. Accepting means that we don't spend time in the past to the exclusion of spending time in the present, and that we don't use our ability to time travel in order to avoid the present. In the next chapter, we'll explore what it might mean to live in the present in the face of grief.

 CHAPTER 9

Being in the Present

During one of my many interviews with bereaved people, I was sitting across a small table from a distinguished older man whose wife had died a couple of years previously. He told me the heartwarming story of their life together, how they had met in high school, gotten married young, had two children and a beautiful home, how happy they were, how much he loved her. He cried a little when he told me about her terminal illness, caring for her in her last weeks, and her eventually passing away. Then he told me that he had recently met a woman who was very different from his wife. She had different interests and was more outgoing, and although dating felt a little strange, he found the time they spent together energizing. He paused, lost in his own thoughts for a moment, and then he said simply, "You know, the thing is, it was good then." Another pause. "And it's good now."

Yearning is not only for the past, for something that was. Yearning also means that there is something we do not like about the present. If yearning were only about the past, we would simply spend some time with our memories, and then switch gears to focus on whatever is happening in the present. But the present moment can be full of pain when we are grieving, which makes the past all the more desirable. If the present has little to say for itself, or if we feel unable to shift our attention and so we do not even know what the present has to offer, the more likely it is that yearning will persist. Beyond the feelings of sadness and anger and amputation that I have already mentioned, pangs of grief can also be full of panic.

Panic

In *A Grief Observed*, the beautiful book C. S. Lewis wrote after the death of his wife, he writes: "No one ever told me that grief felt so like fear." In the worst moments of grief for me, I would have called it panic. After the death of my father, I had no children, I was no longer married, and I had no parents. For the following year I felt completely untethered in the world, without all the usual attachments that had held me in place. The present moment would assail me, often in the evening, and my automatic response was panic. My heart and my mind would race, and I would practically jump out of my chair with restlessness. The only thing that helped me during panic was matching my physical activity to the amount of adrenaline my body pumped out, and so I would go walking fast through my neighborhood, usually in the dark. Eventually the body tires itself out, and the mind does as well, and shedding some tears, I would finally return home.

Neuroscientist Jaak Panksepp concurred with writer C. S. Lewis and with my own personal experience. Panksepp was a pioneer in "affective neuroscience," the field that studies the neural mechanisms of emotion. He insisted that emotion could be scientifically and empirically studied in animals, and he developed a comprehensive model for the range of emotions that the brain produces and the functions of those emotions. One advantage of the warm climate in Tucson is that older academics love to visit, and I had the good fortune to hear Panksepp lecture several times at the University of Arizona, not long before his death in 2017. One of his little-known contributions is to our understanding of the neurobiology of grief. His knowledge was not just academic, as his own teenage daughter died in a car accident caused by a drunk driver.

Panksepp named the neural systems for different emotions with capital letters, like JOY, RAGE, and FEAR. The system that controlled the response to loss, he termed PANIC/GRIEF, highlighting the overlap even in the label itself. Certainly not all aspects of grief feel like panic. Panksepp was referring to (1) acute grief, (2) aspects of grief conserved across species, and (3) grief that has not been elaborated by the higher cortical regions of the brain. He documented that when separated, animals usually undergo a period of increased activity, characterized by increases in heart and respiration rate, the release of stress hormones like cortisol, and distress calls. Panksepp's primary research in this area focused on distress calls, even the ultrasonic ones in some species. He identified what he called the anatomy of grief, or the linked regions of the brain that produced the distress calls when electrically stimulated. The regions include the periaqueductal gray (PAG) in the midbrain, just above the spinal cord. In my second neuroimaging study, the PAG region was activated across bereaved participants when they looked at photos of their deceased loved one compared to a stranger, whether or not they had complicated grief.

Panic, increased activity, and distress calls are likely to bring the separated animal into contact with others of its species, or "conspecifics." One might imagine that the function of PANIC/ GRIEF is to motivate animals, including primates, to come into contact with others. Others of their species could certainly help aid their survival, even if the one who was lost was unable to reunite with their caregiver. Social contact leads to the release of opioids in the distressed animal, which functions both to soothe and to teach. Coming into contact with others is paired with this powerful reward, the internally generated equivalent of opiates, and a powerful reward tends to increase whatever behavior preceded it. How remarkable it would be if we could use this physiological understanding as a unique drug delivery method. A doctor might recommend, "To temporarily relieve your distress, have two conversations with caring people, preferably including a hug, and call me in the morning."

On many of my own panicky occasions, I called my sister or my best friend or, if I could not reach them, another close friend. On other occasions, however, I decided it was too late to call, or I did not feel all that bad, or I had burdened people enough for now. Human beings have the capacity to override all sorts of behavioral patterns that evolution has set in motion. I was lucky to know that these friends would have answered and talked to me regardless of the hour, and their support is very likely what kept me sane. Just knowing I *could* call, even when I did not, was the difference between extreme distress and moderate distress. I am aware of how fortunate I am, though, as there are many people in the world who do not have even a single person they feel they could call in such a situation.

What Does the Present Have to Offer?

If the present moment has only panic and grief to offer, why on earth would we spend time fully mindful of the present? At first, we may be able to bear the painful reality of the present for only a moment. An esteemed colleague in my field once told me that when she was an undergraduate, she got married and they had a baby. Then her husband died unexpectedly. As a single parent with no job and no college degree, she had every reason to feel panic. She told me she knew she could not bear to deal with what that reality meant, but she convinced herself she could probably think about it for two seconds. And that the next day, she could probably bear to think about it for twice that long. And twice that long on the next day. And so on and so on, until she could decide what to do. In fact, she went on to be a very famous researcher and has a wonderful relationship with her adult son. When we allow ourselves the flexibility of mentally time-traveling away from the present, we are trying to protect ourselves from pain, especially when reality is just too painful to bear. Coping this way is very typical in acute grief.

But the present moment also offers us possibility. For example, it offers us other members of our species. And only in the present moment can you feel joy or comfort. You cannot feel those things in the past or in the future. If that sounds unlikely, think of it this way: you can remember times you felt joy or comfort, but you are actually feeling them in the present moment. Memories, or plans for the future, may stimulate you to have these feelings, but the feelings are happening in the here and now. Your body is making cortisol or opioids right now. If you are stuck focusing your awareness in a virtual world where *what if* is true, or where your loved one is alive or your friends understand your grief better, there is a downside: you are missing what is actually happening right now. Although many aspects of what is happening right now may be painful, there are also aspects of the present moment that are wonderful.

Human beings cannot choose to ignore only unpleasant feelings. If you are numb to your momentary experience, you are numb to it all, the good and the bad. You forgo having your heart warmed by the barista who gives you a bright smile or being amused by the puppy loping in the park. If you avoid painful feelings by avoiding the awareness of what is going on around you, what you end up with is being unaware of what is going on around you. It is not possible to avoid only negative feelings. Ignoring the present makes it difficult to learn what works in the new ways you are living your life. On the other hand, when you are present in the moment, the dopamine, opioid, and oxytocin feedback help you move toward a restored, meaningful life.

One year when I was staying with my best friend during the holidays, I was torn between conversations with her and texting with my new boyfriend. She asked me at one point if I had any New Year's resolutions, and she giggled when I told her I hoped to be more mindful in the coming year. I had my phone in my hand while I was saying this, not even looking at her. I was slightly offended by her giggling, as it seemed clear to me that although I was not paying attention to her, I was paying attention to what I was doing. Years later, I came to understand that mindfulness is more than just paying attention. Being in the present moment is awareness beyond your focal point, awareness that includes those who are with you in the *here* and *now*, whether they are friends, cashiers, children, old folks, or strangers. In a way, mindfulness is moving one's attention to awareness of *here*, awareness of *now*, and awareness of *close*. You might be paying attention to what you are doing, but that is not the same as being aware that you are doing it in the present, here in this room, and with the human beings around you. In some ways, I think of this present-moment awareness as wholeheartedness, engaging in what you are doing now in all aspects. This gives you the greatest opportunity to experience what is happening, to see the wonders the world has to offer, and to learn from your interactions with the world.

In the early days of my panicky grief, I did not have the presence of mind to do much at all, let alone learn to change the focus of my awareness. In fact, I kept a note taped to my kitchen cupboard that read, "Cook. Clean. Work. Play." It served two purposes. The note was an intention for what I thought I could actually accomplish during a day, minimal as it seemed. In the moments I found myself overwhelmed or dazed, I could return to this simple list to tell me what to do next. On the days that I did accomplish any aspect of all four goals, I was reminded that this was enough—it had been a good day. Just to be clear, this was normal, typical, average grief I was experiencing, not complicated grief. It took months to remake my life into something I lived fully, and in some ways, it is still a work in progress. In the long term, finding a way to spend more time in the present moment helped me to figure out what that life was like now, and when I knew what life in the present really felt like, I could choose how to spend it.

Insomnia

If grieving didn't make the present unbearable enough, the insomnia that often comes with grief certainly doesn't help. The period following the death of a loved one is a perfect storm that dysregulates all the systems that control our sleep. First, our system is pumping out a combination of adrenaline and cortisol in response to the stress of bereavement, enough to keep anyone awake as though they were drinking extra coffee throughout the day. Combine that with all the changes in what insomnia researchers call the zeitgebers, which means "time givers." Zeitgebers are all the environmental cues that synchronize a person's biological rhythms to Earth's twenty-four-hour cycle of light and dark. Examples of zeitgebers related to falling asleep include eating dinner; a period of quiet prior to going to bed like watching TV or reading; getting into bed with the warmth, smells, and visual cues of your mate; and turning off the lights. Most likely, all of these zeitgebers are disrupted by the absence of your loved one. Each one is instead a cue for grief, a reminder that they are not here. When you are grieving, zeitgebers are not just absent, their absence is also the trigger for grief-related rumination, which maintains our perseverative thoughts and physiological arousal. It's no wonder that we cannot sleep.

Many physicians will prescribe benzodiazepines or sleep medications for bereaved patients, based on patients' desperation when they report sleeplessness. Empirical evidence shows that these pills do not help grief and make the sleep of bereaved people worse over time.[1] Even if you sleep better on a night that you take a sleeping pill, eventually your circadian rhythm gets used to this drug cue. You become synchronized to the feeling of the drug along with the other things you do when you get ready for bed. When you stop taking the drug, you go back to having poor sleep again, or your sleep is even worse. Insomnia rebounds, and you now have to cope with both the absence of your loved one and the absence of a drug your body has come to expect. This is another example of how time does not heal, but rather, experience heals over time. If you take away the experience, even the experience of insomnia, it is harder to learn how to create a life that supports your natural circadian sleep cycle. It is harder to discover what helps your sleep normalize over time.

Insomnia is such an important issue that I want to be very clear—physicians have the best intentions when they prescribe sleeping pills. An accidental finding came from a study of physicians that is relevant here. Researchers wanted to understand why doctors were prescribing benzodiazepines, like diazepam (Valium) and lorazepam (Ativan), to older adults despite all the guidelines against it. The study was not designed to investigate bereavement as a potential prescribing indication, but rather to ask the reason that physicians prescribed these drugs for sleep to anyone. Unexpectedly, more than half (eighteen of thirty-three physicians) spontaneously reported that they prescribed benzodiazepines specifically for acute bereavement.[2] Researchers had not realized how common this was, and at the time this concern was not on the radar of the researchers. In addition, beyond asking physicians, the researchers interviewed fifty older people who were long-term users of benzodiazepines and asked why they had been prescribed the medication in the first place. Twenty percent reported they were initially prescribed these drugs because of bereavement and never discontinued them. On average, they had been taking these drugs for almost nine years. We know that fewer side effects (and effective treatment) would come from learning cognitive-behavioral therapy for insomnia (CBT-I).

Physicians are giving prescriptions to patients because they have empathy for their distress and want to do something. One of the interviewed physicians said, "People who call me up and say my son died, my husband died . . . I would give [benzodiazepines] to them in a flash. Fifteen pills, twenty pills, a month's worth, of course. If this isn't enough, you should make an appointment and come see me. So, they're wonderful drugs for that." I am not suggesting there is never a reason to use these powerful medications. I am suggesting that if the motivation is to provide compassionate care for a patient, but there is no evidence it helps their sleep or grief, the motivation and the prescribing behavior are not in sync.

We cannot force ourselves to sleep, just as we cannot force ourselves to get over our grief. What we can do is provide the opportunities for our natural systems to become regulated again, although even this takes time. We slowly put the pieces of our lives back together and develop new habits, new zeitgebers, a new understanding of what has happened. One way we can help our natural sleep system is to reinforce its regular rhythms. Although we cannot force ourselves to go to sleep, we can force ourselves to get up at the same time every day, the most powerful of the zeitgebers. This waking time resets the whole circadian cycle, and that helps over time. Waking at the same time every day helps even if we feel tired during the day, forcing ourselves to get up with our alarm despite having had very little sleep. In fact, during bereavement, our brain is smart enough to give us what we absolutely need, by taking a slice of each of the stages of sleep. It steals some time from deep sleep, some from rapid eye movement or REM sleep, and some from lighter sleep. That means that although we get less sleep overall, we get all of the parts of sleep we need. This is another amazing example of our brain working on our behalf, on a level we cannot comprehend.

Inserting other cues into the sleep process, beyond medications, is also not a good idea. An older gentleman whose wife died of breast cancer told me he started falling asleep in his big comfy recliner in front of the TV, because he just could not bring himself to get up and face the marital bed. When sleep stole over him late in the evening, he was happy to slip into unconsciousness. But falling asleep in his chair was not a solution—eventually he would wake up with the TV still on, and have to walk the dreaded hall to their bedroom. Without the natural pressure of sleep that comes at the end of the day (because that internal biological impetus was used up while he was in his chair), he would lie in bed awake, feeling sad and lonely, reinforcing the association between their bed and grief. After understanding the biological sleep system better, he made a rule of getting up when the news came on at ten o'clock and preparing for bed, since he often fell asleep in his chair after the headlines. He would brush his teeth during the first news story, and at the first commercial break, he was ready to get into bed. Although he hated facing those same reminders in his bedroom, he would lie down, and the natural slumber narcotic worked more often than not. Over time, he felt less dread going to bed, and felt more confident that not every bedtime would be associated with a wave of grief.

A River of People

There is a poem called "The Sleepless Ones," by Lawrence Tirnauer, that I like a lot. In the poem, Tirnauer writes about being awake in the night, twisting and turning, unhappy about his state. He wonders how many other people are also awake, in this tortured state. If they all got up right now, and came out of their houses to walk in the street, he imagines how a river of people would flow together, all unified by their sleeplessness. It is beautiful.

As it is with insomnia, so it is with grief. Here is the hard thing to wrap our heads around: There is grief in this world—not just yours in particular—and feeling grief at some point is one of the rules of being human. What that allows, on the other hand, is that when we feel grief, we are suddenly joined by hundreds of people who have known grief, from your ancestors to your neighbors to perfect strangers. This river of people may or may not understand you and your particular grief, but they have struggled with grief themselves. You are not alone. As soon as we focus on how grief is manifested in ourselves, as soon as we become fixated on our own experience, we are disconnected from those around us. On the other hand, when we focus on the idea simply that there is grieving, and we are part of it, we find connections. At times we feel ashamed for our strong feelings of grief, or we feel angry at other's reactions to our mood, or we feel weak or disoriented or worried, and on and on. But if we can stop judging ourselves, if we can have compassion for ourselves because we are human and because this human life comes with grief, we may find it easier to connect with others as well.

This is an aspect of closeness, a dimension that the brain uses. Just as you can shift your mind from the past to the present, could you shift your mind from feeling distant to feeling close? Consider how similar you are to someone you know. You both have frustrations. You both hope for happiness. You are both tied to a physical body that has aches and pains. The content of these similarities may differ, but the human experience overlaps. Think back to that row of overlapping circles from chapter 2, the Inclusion of Other in Self scale. Perhaps if you move two circles around, like they are planets in a model of the solar system, what you see would change. Through moving how you are lined up to look at them, two globes that do not even touch might come to share some space by changing your perspective. Perhaps you and another person could be considered close, from another perspective.

A few years ago, I drove to Wyoming to see the solar eclipse, a spectacular event that happened in the middle of the day. For a brief time, I could see that the moon can block the space between

on, many grieving people are unable to do much productively, as our mind, our brain, and our body are too dysregulated to function properly without our loved one. But over time, we have an opportunity to learn how to respond to each moment as it presents itself. We can consider what is in our best interest, the pros and cons of spending the present yearning for the past. We may be avoiding what is happening in the present moment, not engaging in what can be seen, felt, and tasted right now. Or we may simply be unaware of where our mind is, in the habit of mind wandering unless our attention is grabbed by something, or unless we are doing a task that requires focus. It is harder than it seems to move our attention. It requires effort, especially at first. Because our brains generate thoughts at a persistent rate, we are not likely to stay in the present for very long. But repeating this skill over and over will actually make changes in our brain. When people practice new ways of thinking—from learning meditation to undergoing psychotherapy—neuroimaging studies show their patterns of brain activation change. It is a remarkable idea, that the content of our thoughts, or where we deploy our attention, changes the hard drive of the brain, the wiring of our synapses. This is a dynamic process. Our neural connections generate the content of our thoughts, and at the same time, our guiding the content of our thoughts changes those exact same neural connections.

I am reminded of an analogy made by a friend of mine who is a massage therapist. She told me she believes that her job is not just to mechanically decrease tension in muscles. The key is also to bring the client's attention to specific places in the body, in order to enable them to relax their own muscles. Her role is to guide attention; the change is actually being made internally by the client. What can we use to remind us to shift our attention to the present?

One way to explicitly notice we are in the present while our thoughts are turned to the one we have lost makes use of memorials. Memorials can be single events, but in many cultures, there are daily or weekly rituals to connect our outer behavior with our inner thoughts of our loved one. Lighting a candle is a very common example—the action of striking a match and watching the flare, the smell of the smoke and the candle wax, the mental notation of our present activity, coupled with the thought of our relative or friend—all these remind us that while we are in the present, we are always incorporating aspects of our past.

Other rituals are less obvious. Many years ago, our cat died. This was my first long-term relationship with an animal, my first grief for that special kind of relationship. After he died, I began buying flowers. This had not been possible while he was alive, because he would inevitably find them, eat them, and then throw up all over the house. For a long time, I could not figure out why it was important to me to keep buying flowers. My motivation seemed even stranger, even to myself, because looking at the flowers was a little painful, since they brought his absence to mind. But I also enjoyed the flowers, with their delicate petals and beautiful smell. Eventually I realized I loved having my kitty in my life, but that did not mean I did not miss having flowers in my home during his lifetime. In the present, I enjoyed having flowers, even though they were a reminder that he was gone. It is not a simple trade-off; I did not get to choose between these two, as though they were options. It was just the reality of the present moment in which I found myself. There are always some aspects of the way things are that I enjoy and others that I do not. I cannot pretend things were only good when my sweet cat was alive. Buying flowers was a way to remind myself that I am here now, and I want to really be a part of now, with flowers and memories of him and all of it together.

Mind-Wandering Thoughts

Neuroscientist Noam Schneck at Columbia University published several papers in the late 2010s, tackling some of the difficult problems in understanding how grief is processed by the brain. Schneck employs neural decoding, a new technique in neuroscience. This method uses highly sophisticated algorithms to look for "fingerprints" in brain activity that occur when we have a thought about something specific. Here is how it works. Schneck asks the participants to think about their deceased loved one while they are being scanned. He helps participants produce these thoughts by showing them reminders of the deceased, including pictures and stories. We will call this the photo/stories task. Participants also view stories and photos of a stranger, like the control condition we have seen in previous studies. After the scan, a computer identifies the brain activation patterns unique to thoughts of the deceased, or the fingerprint of deceased-related thought, compared to the thoughts activated by the stranger. Because these patterns are being found by a computer, the technique is called machine learning. More specifically, machine learning is when the computer "learns" to identify thought content by looking for patterns in one set of data. Then the computer is "tested" by seeing whether it can use that same pattern in a different set of data to accurately predict the same thought content. In Schneck's study, the brain activation pattern, the neural fingerprint of deceased-related thoughts, included activation in brain regions we have found before in grief studies. These included the basal ganglia, the neighborhood in which you can find the nucleus accumbens.

The amazing thing about this process of machine learning is that once Schneck had identified the neural fingerprint of deceased-related thoughts, he was then able to use that same fingerprint to look for thoughts about the deceased during a different neuroimaging task. Participants also did a sustained attention task, a task so boring that it usually leads to mind wandering. They lie in a scanner for ten minutes, pressing a button every time a number appears, unless it is the number 3. As you might imagine, this is not a very absorbing activity, and soon participants' minds turn to other thoughts, as the researchers expected. Every thirty seconds or so, these participants were asked if they were thinking about their deceased loved one.

Schneck and his colleagues wanted to know if the neural fingerprint identified in the photos/stories task could accurately predict when participants were thinking about their deceased loved one during the sustained attention task. Sure enough, the neural signature that the machine learning algorithm produced from the first task was able to predict with greater-than-chance accuracy when participants said they were thinking about the deceased in the second task.

Before you decide this is too creepy, or that neuroscientists are trying to be mind-readers, remember that there is no way to find the neural fingerprints of thoughts without a person's permission. The person has to tell you when they are thinking about a particular thing in order to create a data set the computer can learn from, which requires the participants' willing collaboration. And the neural decoding, while impressive, is not even close to 100 percent accurate. Thoughts are conscious experiences, and the neural fingerprints of those thoughts can only be learned by a computer if there are lots of reports from the person. No researcher can figure out what someone is thinking unless the participant is actively trying to help them match up what they are thinking about in the moment with maps of brain activation.

So, how often were bereaved people's thoughts focused on the task in the present moment? The results of Schneck's neuroimaging study revealed that during the sustained attention task (when people's minds were actually often wandering), 30 percent of the time they were thinking about their deceased loved one. In real life, during the early days of grief, trying to get a task done is often interrupted by intrusive thoughts about the deceased loved one. Here's the most interesting result from this study: the more times the neural fingerprint of deceased-related thought showed up in participants' brain activity, the more often they avoided thinking about the deceased or their grief in everyday life. So, the more they tried to avoid thinking about the person, the more they thought about them unintentionally during mind wandering. From this we see that while cognitive avoidance can be a strategy that bereaved people use to get relief from frequent, painful thoughts of loss, higher avoidance also goes along with a higher number of intrusive thoughts. Suppressing one's thoughts is, ironically, related to a rebound of those thoughts. We need to discover new strategies to help bereaved people manage their painful thoughts in the present moment, since avoidance does not help them very much in the long run.

Unconscious Processing of the Loss

This first study that Schneck conducted focused on conscious, reportable thoughts of the deceased, even when they occurred in the midst of trying to do something else. The second study Schneck did was even more interesting. He wanted to understand more about the unconscious processing of the loss. For thoughts that are conscious, he could just ask people what they were thinking about. To study unconscious processing, he had to find a way to look for a neural fingerprint that did not rely on reports from people. Unconscious processing is similar to what we considered in chapter 1: the brain learns about the absence of your loved one through experiencing your new world over time and with experience. Let's say you realize that you are no longer opening your husband's sock drawer after you do the laundry; this new behavior has developed because of lots of background processing of repeated experiences. We do not always need to be engaged in grief work or deliberately focusing on the loss, because the brain is learning and adapting even when we are not explicitly aware of it. A graduate student working with me, Saren Seeley, compares this to the way a computer runs programs in the background when you are typing a document on the screen. Those invisible background programs are making it possible to do the task at hand. However, there is a limit to how many resources a computer can allocate to those background programs before the task you are trying to do grinds to a halt.

Schneck looked for a neural fingerprint for unconscious processing of the loss by observing when participants in the second study were slowed down by reminders of the deceased. I'm sure you've noticed how many things in your environment remind you of your loved one when you are grieving, and that those reminders distract you. Scheck's neural decoder compared the fingerprint of the brain distracted by deceased-related words in a reaction time task, compared to the faster processing of other words. The computer set to work hunting for patterns of brain activation that distinguished that difference in selective attention. In this second study, the computer was not trying to find specific thoughts about the deceased with its algorithms, it was just trying to find the slowing of reaction time when the brain was paying attention to words related to the deceased. Here is the punchline: more slowing, or more unconscious processing of the loss when doing other tasks, was linked to reports of fewer and less-intense grief symptoms. More neural fingerprints of this unconscious incubation were linked to better adaptation. We don't have any control over our unconscious thoughts, but it's interesting that this is how it works! To summarize, what Schneck found in the two studies was that conscious, intrusive thoughts about the deceased were linked to more grief. Avoiding those thoughts was associated with their happening more often. On the other hand, unconscious processing was associated with less grief. So, while the conscious thoughts distracting you may not be helpful (although possibly unavoidable), the unconscious thoughts during mind wandering do seem helpful.

Bereaved people who use avoidance seem to be screening their unconscious mental processing to stop thoughts of their deceased loved one from breaking into their conscious awareness. Schneck compares this to using an inefficient pop-up blocker. Screening our incoming thoughts works, to some extent, and blocks the pop-ups at first. But over time the system gets overloaded and ultimately the pop-ups get through. Bereavement science has a long way to go to understand the relationship between conscious and unconscious processing of grief. Many more studies need to be done to understand the way that both avoidance and rumination can lead to, or maintain, prolonged grief disorder. But the investment of smart, young neuroscientists in the neurobiology of grief encourages me to think that we are on the trail of discovery.

Love

After a loved one dies, they are clearly no longer with us in the physical world, which each day proves to us. On the other hand, they are not gone, because they are with us in our brain and in our mind. The physical makeup of our brain—the structure of our neurons—has been changed by them. In this sense, you could say that a piece of them physically lives on. That piece is the neural connections protected within our skull, and these neural connections survive in physical form even after a loved one's death. So, they are not entirely "out there," and they are not entirely "in here," either. You are not one, not two. That is because the love between two people, that unmistakable but usually indescribable property, occurs *between* two people. Once we have known love, we can bring it into our awareness, we can feel it emerge and emanate from us. This experience reaches beyond the love for the flesh and bones of the person we once knew on this earthly plane. Now loving is an attribute of us, regardless of who we share it with, regardless of what is given to us in return. This is a transcendent experience, a felt sense of being loving without needing anything in return. In the very best moments together, we learned to love and to be loved. Because of our bonded experience, that loved one and that loving are a part of us now, to call up and act on as we see fit in the present and the future.

CHAPTER 10

Mapping the Future

On a Friday in 2002, two-year-old Ben was at home with his mom, Jeannette Maré, his older brother, and a friend. Ben's airway swelled shut and, despite all efforts, that Friday became the much-too-soon last day of Ben's life. Jeannette says the pain was indescribable as she and her family lived their new reality. They began to work with clay as a way of coping and created, along with friends, hundreds of ceramic bells in their garage. On the anniversary of Ben's death, they hung these bells randomly all over Tucson, with written messages to take one home and pass on the kindness.

Jeannette says she realized that getting through the day was made possible by her community, her dear friends. She wanted to find a way to pass on that kindness, to help others who needed it. From this tragic situation, Ben's Bells was born, a nonprofit with a mission "to teach individuals and communities about the positive impacts of intentional kindness and to inspire people to practice kindness as a way of life." Ben's Bells now teaches intentional kindness programs, from kindergarten to college. The effect has been remarkable. Passing any school in Tucson, one sees a green tile mural that says "Be Kind." All over town, cars carry the signature green bumper stickers shaped like a flower with "Be Kind" written in the center. Giving or receiving one of the handmade bells, topped with a ceramic flower, is a sacred act.

Ben's Bells has been so impactful because it was born out of a very real truth that can happen in grieving. Not everything that people said to Jeannette was kind or helpful. Often their words were hurtful, even with the best of intentions. I spend my life thinking about grieving and yet I still cringe when reflecting on things I've said to a grieving person. It's hard to know what to say and we so often get it wrong.

Jeannette has a background in communication, and her training helped her to see that we need to talk about *how* to be kind. What feels "kind" to a grieving person requires awareness of what grief is like, and Jeannette doesn't shy away from difficult conversations, from honest explanation of what grief feels like. The grieving person may be sad, or angry, and that is the natural response to loss. For those around them, cheering them up is not the goal. Being with them is the goal. Jeannette also realized it was what the words conveyed, even more than the words themselves, that mattered. She wanted to help people to understand that really listening to what a bereaved person is feeling and where they are that day is important.

Even saying that you don't know what to say to them, but that you love them and will be with them through this, is vulnerable and powerful. The practice of giving a gift, such as a bell, creates an opportunity to reflect on how to give, how to be present, how to be kind. Because of Jeannette's experience with grief, and her honesty about her own experience, she transformed both the painful and the supportive experiences she had into a program that enables us all to benefit from Ben's life, even though we did not know him. Ben's life has touched many, many people. It is not the life Jeannette imagined living, but she lives a life restored.

Grief and Grieving

As I described in this book's introduction, grief is different from grieving. Grief is the painful emotional state that naturally rises and fades away, again and again. People might imagine that grief is "over" when the waves happen less often, or less intensely. They are right in one sense: if the goal is to suffer less-intense and less-frequent pangs of grief, this reduction is likely to happen naturally over time with experience. On the other hand, if a bereaved person does not experience the lessening of intensity and frequency over time as they were anticipating, they may begin to ruminate not just on their loss but on their reaction to it as well. They may begin to wonder, Is my grief normal? Others are expecting me to "move on" and I do not feel as though I have "moved on." Does that mean that I will always feel this way? Monitoring like this has the unfortunate effect of keeping grief at the forefront of your mind, which can heighten and prolong your grief reaction instead of allowing it to gradually become less painful with time.

On the other hand, I think most bereaved people hope for something more than just a decrease in the intensity and frequency of pangs of grief when they think about grieving being "over." Restoring a fulfilling life may be a better definition, pointing to adaptation, which I think is more accurate than thinking of grief as being "over." And a meaningful life involves a lot more than simply the end of frequent, intense pangs of grief. If one believes that the only way to have a meaningful life is to be with the person who has died, this goal can never be reached. Instead, one may have to give up this specific way of achieving the goal of a meaningful life, while elaborating on other ways. Let's face it, that is just plain hard.

You have a better chance of reaching your goal if you have many ways you might consider your life meaningful. This requires great courage and flexibility. It requires your brain to learn new things, aided by paying attention to what you actually find meaningful and satisfying in the present moment. But this shift can also lead to a life of love, freedom, and contentment, albeit a different life than you had before. Grieving is the change from having your attachment needs fulfilled by your deceased loved one, to having them consistently fulfilled in other ways. That does not necessarily mean fulfilled by one other person. Having a meaningful life is not the same as remarrying or having another child. In fact, these relationships might distract you from pursuing a meaningful life, if they get in the way of reaching your goal.

Moreover, what constitutes a meaningful life has very likely been changed by your recent close acquaintance with mortality. Death has a brutal way of clarifying to us what is meaningful. This clarity can lead to the discovery that our day-to-day activities are completely unrelated to the values we hold. Such a realization is frustrating and depressing and can lead to great upheaval if we are willing to change our day-to-day lives in pursuit of the newly realized values. We may be less willing to listen to a colleague go on about her life's drama, if doing so feels fake and meaningless. We may not care as much about proper etiquette at a family event, in light of recent events. This discovery of the mismatch between our values and day-to-day minutiae may lead us to feel annoyance at the situations we find ourselves in, or it may make us feel fearless in expressing strong emotions or pursuing new goals. But we don't live in a vacuum. These emotions or changes in us are not easy for our living loved ones to adjust to either, and we may find friction with them as a result of our new awareness and changed priorities. Some bereaved people come to find that all the people in their address book have changed. During grieving, we are sometimes redefining our identity, based on what the brain is learning about our new world and what we enjoy or find worthwhile. If our identity is an overlapping circle with someone who is no longer there, should we find it so surprising that we change without their constant influence, or that we need to redefine and update our pursuits and circumstances?

What Is the Plan?

The ability to imagine our future, a new and unknown future that no longer includes our deceased loved one, seems to use a similar brain network as remembering our past. That may sound odd, but Canadian cognitive neuroscientist Edward Tulving showed that our ability to travel in time, both forward and backward, shares some important characteristics. As we've discussed in earlier chapters, memories are what happen when our brain replays neural activity that was generated during the original event. This creates a perception of the event, a memory, with the knowledge that it is being recalled in the present. Imagining the future is also a recombination of possible pieces of an episode, with the knowledge that they might happen in the future. For the virtual projection into the future to be plausible, the brain relies on things you have already experienced and could experience again, combining them in novel ways.

A while back, I went to Las Vegas to celebrate the fiftieth birthday of one of my friends. I remember what my hotel room looked like, and can picture walking from the window, past the bed, and into the large bathroom. I remember the amazing flavor of a shake I drank and the visual spectacle of a Cirque du Soleil show we attended. I recall what I wore to my friend's birthday dinner and unpacking those clothes in the hotel room. This helps me to imagine a potential vacation I would like to take in the future. I may consider what size of hotel room I would like to book, and whether I would want a window that faces downtown. I might make a reservation for a restaurant that serves creamy shakes. I may think about what shows I would like to see, and anticipate what my friends will also find entertaining, like visual spectaculars as opposed to lounge singers. Planning for packing, I might mentally try on several outfits and consider what would fit for the climate, the season, and the activities I will do. Considered in this way, you can see the similarities in the process of recalling a memory and imagining an event in the future.

Neuroscientists have discovered two pieces of compelling evidence for the idea that retrospection and prospection share neural machinery. First, when people's brains are scanned as they are remembering their past and imagining their future, there is significant overlap in the brain regions used for these two mental functions. Second, when people have difficulty remembering events in the past that happened to them, they also tend to have difficulty imagining the future and what they might do.

Understanding how the brain works without key brain regions intact can teach us about how the brain functions in people with normal memory as well. Tulving studied a famous patient named K.C., who had a deficit in the capacity for both autobiographical past and future thought. K.C. had suffered a head injury in a motorcycle accident, which had very specific consequences for his mental functioning. He retained his intelligence, his ability to shift his attention, and his language skills. He had normal short-term memory, meaning he could remember something recently shown to him. His general knowledge of the world, the form of knowledge called semantic knowledge, was also good. He could identify a car he had owned, his childhood home, and members of his family. However, he could not remember a single experience associated with any of these items or people. He knew they belonged to him without being able to describe any memory that included them. Tulving also assessed K.C.'s ability to consider his future. If he asked what K.C. would do tomorrow, K.C. was unable to answer the question. He reported that he did not know and described his mind as blank, similar to the way his mind was blank when he tried to think of events that had happened in his past. Remembering one's past and imagining one's future uses the same neural machinery, and that aspect of K.C.'s brain was damaged, leading to deficits in his ability to do both.

Part of the Past, Part of the Future

The capacity to remember the past and to imagine the future has specific application to people with complicated grief. When Harvard psychologists Don Robinaugh and Richard McNally tested bereaved people's ability to recall personal memories, they found that those who have the most difficulty with grief also have difficulty remembering specific details about their own past, unless the memories include the deceased loved one. Similarly, they have difficulty imagining details of future events, unless they imagine a counterfactual future in which they envision events as though the deceased were still alive.

In order to determine this, Robinaugh and McNally asked a group of bereaved people who were adjusting resiliently and a group with complicated grief to bring to mind four situations in as much detail as possible. They explained to the participants the difference between recalling more general events and specific, autobiographical episodes. General events would include those that occurred over a long period, like the summer after high school; events that occurred regularly, like high school biology class; and general knowledge about one's past, like the name of one's high school. Specific episodic memories would include details about an event such as one's high school graduation ceremony. These different types of memory are stored differently in the brain. Each participant was asked to remember or imagine an event in response to cues like successful, happy, hurt, or sorry—half of them with the deceased and half without. People who were adjusting resiliently showed no difference between the ability to generate a specific memory for the past or imagine an event in the future, regardless of whether the event included the deceased or not. Those with complicated grief, however, generated fewer specific remembered or imagined events if they did not include the deceased.

Robinaugh and McNally also tested the working memory of the participants. This ability, being able to hold things in mind, is necessary for both remembering and imagining. People with complicated grief are more likely to remember specific events with the deceased because when asked, if the deceased has been on your mind a lot, those are the memories that get reported. When asked to think of a time without the deceased, it may take great effort to come up with any that did not include them. In fact, the working memory test bore this out. The fewest specific memories without the deceased were generated by those with complicated grief and poorer working memory, presumably because coming up with memories that do not include the deceased requires more effort for them.

Why would those with complicated grief have more memories with the deceased, and even more strange, why are future events more easily imagined with the deceased? There are at least two possible reasons. One is that if we are often ruminating about the deceased, the ingredients that constitute a memory are more likely to include the deceased, and therefore to be accessible when we are asked to report one. The other reason is that if our own identity overlaps with the deceased, such as thinking of ourselves as "wife," then imagining ourselves in the past or the future is more likely to include the deceased person as well. If the very nature of our self implies that we have a husband, then imagining ourselves in the future automatically brings him, too. And it's easy to see why we would feel as though part of our self is missing after our husband's death, if our identity integrates "wife" as part of "self." On the other hand, if we have many aspects of our identity that are unrelated to the deceased, such as "sister" or "supervisor," then events that come to mind are just as likely not to include them.

Restoration

Restoring a meaningful life is half of the dual process model of coping with bereavement. To restore a meaningful life, we have to be able to imagine that life. The inability to generate possible future events is at the heart of hopelessness. We have to be able to imagine the future sufficiently enough at least to make plans, even if only for next weekend. I hear often from widowed older adults that evenings and weekends are the worst times for them, when they feel the loneliest, because everyone else has things to do and people to do them with.

If grieving is a kind of learning, that means we can learn on Saturday and Sunday how good our planning for the weekend was. We can assess whether we actually enjoyed our plans and found them meaningful, whether they enabled a productive week after.

During bereavement, this is a process of trial and error. We make a plan, but we cannot entirely imagine how it will go, now that we are widowed or orphaned and feeling estranged from people around us. Luckily, we do have life experience and we have some intuition. No, I do not want to go to a rock concert until the wee hours of the morning. Yes, I need to see someone over the weekend or I will feel very lonely and depressed. But does that mean taking a road trip to see an acquaintance? Or would I rather spend time with a friend, having coffee? These choices may be less clear. Nonetheless, if we make a plan and carry it out despite our uncertainty, we get feedback. While I was grieving, I learned I had better grocery-shop first thing in the morning on Saturday, because generally I had very little motivation to do so and little appetite, and if I left the task until later, I would end up eating cereal all week.

Restoration is even more important when imagining upcoming holidays. Holidays are a notoriously difficult time for people who are grieving, because the ritual nature of holiday events brings memories to mind, and the social nature of them emphasizes the absence of those we used to celebrate with. Planning for the holidays means you must imagine yourself without your loved one, and many who are grieving avoid even thinking about holiday plans. My mother died on December 31, and my wonderful in-laws asked me, my sister, and my father all to come and stay with them in Texas for the following Christmas. None of us could quite imagine what that would be like exactly, but we thought we wanted to be in a place with fewer reminders, at least for the first year. (The first year in particular means a lot of trial and error.) In this case, going to my in-laws was a good choice for my family. The key is to figure out what worked and what did not work, so this knowledge can be applied to the next holiday season. And the next, and the next, because the holidays will just keep happening, year after year. Of course, one also has to keep in mind that how you are doing in the first year of grieving, and how your family is doing, is different from how you will be doing in the second year of grieving, and the same rules may not apply. The good news is, if we are paying attention to the present, remembering last year, and planning intentionally, we can get better at having meaningful holidays and new experiences—not necessarily always joyful, but at least meaningful. Even if it all turns out less enjoyable than you hoped, there was a reason, an intention behind what you did—you are trying; you're out in the world, learning how to carry the other person inside; learning to listen to others, not just the voices in your head; and you're making new memories, testing yourself in new experiences (and surviving).

The Future of Our Relationship

We live our changed future each day, and our identity changes as we survive and eventually even thrive following our bereavement experience. Is it possible, then, that our relationship to our deceased loved one changes as well? For more than a decade after her death, I would say my relationship with my mother stayed pretty much the same. I felt alternately overwhelmingly guilty for not being a better daughter and not helping her feel better day-to-day, angry about the way she had raised me, and depressed about what it all meant for how my life would play out. I thought of myself as a product of her genes, her controlling child-rearing, and my own relentless need to solve everyone's suffering. Strong emotional reactions require great skill to manage, and in my twenties and thirties I did not have those skills. Then, for a long time, these feelings declined in intensity, although I would say they continued to inform how I viewed the world.

I watched my friends as they also hit their forties, and for some of them, becoming professionals, becoming mothers or fathers, and gaining life experience changed their relationships with their living mothers. I saw my friends become more compassionate toward their mothers' moods and idiosyncrasies. I saw them become more grateful for the sacrifices their mothers made to give them an education, self-esteem, or a stable home. I experienced grief for the first time in a new way—this was not something I was ever going to have with my mother. We could never have a transformed relationship, as two adult women. The end of her life took that opportunity away, a loss of our potential relationship I could never have anticipated when I was in my mid-twenties. Suddenly the relief I had experienced because of her death, because I no longer had to deal with the difficult interactions she created in my life, was replaced by grief over what could have been.

I realized that along with the fresh grief, I was also more grateful for the things my mother had given me. There was no way I would have survived academia if my mother had not insisted on the discipline of practicing the piano every single day and seeing the long-term improvement that comes from incremental hard work. I would not have navigated the social world as well as I have without having been trained in the cultural standards of thank-you notes, appropriate footwear, and how to make small talk, despite

the fact I despised this training. I realized my mother was interested in any skills that could give me an advantage in this world,
and she was willing to make sacrifices to make sure I learned them.
I thought more about her feminist principles, instilling in my sister
and me the idea that we could achieve anything we set our minds
to. I thought of her capacity to give us her fullest attention, and to
speak to us as curious and intelligent beings even as children, when
other parents did not always seem to show this same level of interest. I was suddenly able to fondly remember long-forgotten specific
memories of her physical affection for me when I was little, though
I retreated from these interactions as a teenager and young adult.

Somehow, I came to believe that if she were no longer con-
strained by her human form, on this worldly plane, she would be
the best parts of herself all the time. At some point it seemed I
could take those best aspects of her going forward in my own life.
It was not that I had failed to grieve for her previously, that I had
denied my feelings and they were now coming up. It was just that
as I aged, the dual process model of coping with bereavement continued to apply. While I felt grief over her absence in this new part
of my life, I continued to adapt to her death, and I continued to
learn how to restore a meaningful life. My relationship with her,
present and past, was transformed when I focused on all the good
she wanted for me, despite all the difficulties we'd had throughout
our relationship.

Our understanding of ourselves changes as we gain wisdom
through experience. Our relationships with our living loved ones
can grow more compassionate and resonant with gratitude as we
age. We can also allow our interactions with our beloved ones who
are gone to grow and change, even if only in our minds. This trans-

formation of our relationship with them can affect our capacity to live fully in the present, and to create aspirations for a meaningful future. It can also help us to feel more connected to them, to the best parts of them. It can allow us to become the best daughter, son, friend, spouse, or parent they would have wanted us to be if they had lived to see it. Our love for them is still there, but we must find a different way to express it, a different outlet for our love for them. Although they can no longer benefit from our kindness and care directly, their absence from our physical world does not make our relationship to them any less valuable.

New Roles, New Relationships

The restoration of a meaningful life very often means developing a new relationship or strengthening an attachment with someone we already know. Bringing someone new into your life can lead to an eruption of grief, even after a period of relative calm. In the enjoyment of a new relationship, the mere presence of the new person can be a reminder of the absence of your deceased loved one. This requires time and gentleness with yourself, and remembering that the new person you love now and the person you loved then are not the same. Gaining a loving, supportive relationship does not mean forgetting or rejecting the one that came before. A new relationship is ripe with new things to be learned, and many adjustments have to be made in order to be present in the current relationship and not to live in the virtual reality of the previous one. For those who are supporting someone who is grieving, there is real benefit in listening and offering encouragement, without judging when it is "normal" to develop new relationships.

One reason we may question a new relationship has nothing to do with whether it is good for us, or fulfilling, or enjoyable. Psychologists Amos Tversky and Daniel Kahneman (winner of a Nobel Memorial Prize in Economic Sciences in 2002) demonstrated that human beings find losses to be twice as powerful as gains. This is called loss aversion, and although I have not seen it applied in the context of bereavement, I think the concept may help us to understand the common experience of having misgivings about a new relationship. If we decide we are ready to date, for example, or to go on a trip with a recently acquired friend, the time spent with the new person may not be very satisfying. Or more accurately, it may not be as satisfying as the time we spent with our deceased loved one. We may not feel as good as we had hoped we would. We expect to feel good because we are exploring a new relationship, and a new relationship is supposed to be fun and exciting. We may expect to feel less grief, because we have chosen to do something new in our lives, after a period of mourning. Note the significantly high bar those two expectations demand, however. If losses are psychologically twice as powerful as gains, then we would have to feel twice as good in a new relationship as we did in our previous relationship in order to feel the same level of happiness. Gaining a new relationship is simply not going to fill the hole that exists. Here is the key—the point of new roles and new relationships is not to fill the hole. Expecting that they will can only lead to disappointment.

The point is that if we are living in the present, we need to have someone who loves us and cares for us, and we need someone to love and care for as well. The only way to enjoy a fulfilling relationship in the future, however, is to start one in the present. If we can imagine a future in which we are loved, then we must start a relationship that eventually will become important to us in a way that is different from our previous relationship, but rewarding and sustaining. This is why attachment relationships, our loved ones, are different from other social relationships. If our boss quits, or we no longer see a teacher after a class ends, there is another person who can fill that role. We share a deep commitment with our partner, our child, our parent, our best friend. If an attachment figure is lost, then the great trust invested in that person over many years and through many shared adventures is lost as well. There will not be another person available who can easily fill that role. A huge investment must be made again. In order to develop another strong bond, great trust must be built over time and through shared experiences. But it will happen if we start now.

Flying the Coop

This reattachment aspect of later bereavement adaptation could be compared to another period of our lives when it is normal for us to transition from one important relationship to another. As adolescents, we must learn to rely less on our parents, to go out and explore the world to find a new relationship. We look for a partner who will become the central person in our lives, the person who meets our attachment needs. Most people recognize that although this is a normal and necessary experience, leaving the nest is also extremely stressful. Different people take different amounts of time to successfully fly the coop, and this period can be fraught with hazards and setbacks. Although it is a normal, albeit stressful process, it might also co-occur with mental health complications, such as depression, drug overuse, anxiety, and even suicidal thoughts. Like leaving home, bereavement is a normal process that is difficult, and also a time ripe for mental health issues to arise, and those issues may be aided by professional help. In many ways, I think of the transition from nurturing parents to a romantic partner as similar to the necessary reattachment that happens when a bereaved person finds a new love interest, or a new best friend, after the death of a spouse.

There are some key differences, of course. When we are leaving home, most of our peers are going through the same transition, and so we usually have built-in support among our friends. Leaving home is also quite predictable in terms of approximately when it will occur. Many societal systems are in place to aid in this transition, from undergraduate dorms, to basic training in the military, to the year spent on a mission in some religious traditions. In contrast, the death of a spouse happens only to some people, and can happen at any time during the lifespan. In our bodies as well, coming of age and leaving home coincide with a specific transitional period. The hormones that motivate us to take risks, to explore the world, and to have sex are in full effect. Because bereavement often occurs at older ages, we must seek out new relationships and new roles without the benefit of high levels of motivating hormones, due to normal aging.

A final difference is that leaving home does not mean your parents disappear from your life. Our parents retain an important role afterward. This is sometimes referred to as the attachment hierarchy, where eventually a spouse may become the most central figure at the top of our pyramid of loved ones, but parents are often still present and offer important sources of comfort for us at lower levels of the hierarchy. Rather than thinking of a hole created in the pyramid when a loved one dies, a different way to conceptualize bereavement is that a continuing bond, the mental representation of our deceased loved one, might still appear in the hierarchy. Because the deceased person is not able to fulfill our earthly attachment needs, however, our relationship to another person, or other people, rises in importance. Allowing someone new to become important is good, and healthy, and maintaining a mental or spiritual bond with our deceased loved one can also continue on a different level of the pyramid.

When explaining who an attachment figure is, I ask two questions. First, does this person think I am special, and do I think they are special, compared to other people in the world? Second, do I trust this person would be there for me if I needed them, and do I trust I would make the effort to be there for them, if they needed me? If a relationship fulfills those two questions, regardless of what the social role of the person is, then likely one's attachment needs are being met. This could be a neighbor, a sibling, a secretary, a pet, or a partner. What society calls them is much less important than the role they play in your life.

When Did You Start Loving Them?

Our loved ones not being with us is a continuation of them being there, as exhaling is a continuation of inhaling. The fact they are not there affects us, affects our lives, our decisions, our values, as much as their being there does. Holding your breath is not the same as never having breathed. So, too, your life in the beloved's absence after their death is not the same as it would have been if they had never lived. I sometimes ask, When did your relationship start? Was it when you got married? When you first kissed them? When you first saw them? By the same token, when do they stop being a part of us? When they are out of our sight? When they die? When we bury them? When we love other people? When we move away from the home we shared together? These are all parts of knowing them, of being affected by them, and loving them, and those never end.

As important as it is to study those who are having the greatest difficulty adjusting to life after loss, there might be much to be gained by studying the people who have created beautiful, meaningful, loving lives after terrible losses. Although this resilience has not yet been the subject of investigation in neuroscience, in psychology it is called post-traumatic growth. People who have experienced enormous growth have much to teach us, and their brain may have an important role, from how they process reminders of their loved one to the ways they become loving, compassionate, and effective in their present lives.

CHAPTER 11

Teaching What You Have Learned

Now you know that grieving is a form of learning. Acute grief insists that we learn new habits, since our old habits automatically involved our loved one. Each day after their death, our brain is changed by our new reality, much as the rodents' neurons had to learn to stop firing when the blue LEGO tower was removed from their box. Our little gray computer must update its predictions, as we can no longer expect our loved one to arrive home from work at six o'clock, or to pick up their cell phone when we call them with news. We learn that our loved one does not exist in the three dimensions of *here, now,* and *close* that we are expecting. We find new ways to express our continuing bonds, transforming what *close* looks like, because while our loved one remains in the epigenetics of our DNA and in our memories, we can no longer express our caring for them in the physical world or seek out their soothing touch.

Although we may still talk to them and live in ways that would make them proud, we must do these things while we are aware that we are in the present moment. Instead of imagining an alternate *what if* reality, we must learn to be connected to them with our feet planted firmly in the present moment. This transformed relationship is dynamic, ever-changing, in the way that any loving relationship is ever-changing across months and years. Our relationship with our deceased loved one must reflect who we are now, with the experience, and perhaps even the wisdom, we have gained through grieving. We must learn to restore a meaningful life.

When I say that grief is a kind of learning, I don't mean learning something easy. This is not like mastering a specific skill such as riding a bike, learning how to keep our balance and how to use the brakes. This type of learning is like traveling to an alien planet and learning that the air cannot be breathed, and therefore you need to remember to wear oxygen all the time. Or that the day has thirty-two hours, even though your body continues operating as though it has twenty-four. Grief changes the rules of the game, rules that you thought you knew and had been using until this point.

Because the brain is designed for learning, thinking about grief from the perspective of the brain can help us understand why and how grief happens. The brain has multiple information streams that we can raise to consciousness. We can experience the yearning for our loved one, the desire to search for them, the belief that they will return. This is ingrained in us through evolution, through epigenetics, through the habit of being together. We also have memories of the deceased person, memories of their death or of learning the news of their death, memories of all the events in the first year of grieving, the first time we did each thing in the absence of our beloved. We can bring these to mind as well. Finally, we can shift our attention to the present moment, which can be so vibrant and full of possibility. We can rest in just this moment, in just this. Nothing else. When we give ourselves a moment to rest and give our brain a chance to practice what it is like to be simply mindful of our surroundings, that particular state of mind or pattern of neural connections can be reached anytime, anywhere. This state of mindfulness is not better than the reveries of fond memories or the yearning state that exemplifies our bond. But the skillful ability to shift when we need a break, even if only for a moment, can help us endure the unbearable reality of loss. If we gift ourselves this moment, we may find opportunity in the present, even when we least expect it. If we are aware of the present, and can acknowledge its value, that opportunity for connection or joy will not pass us by without our notice.

What Science Knows About Learning

Decades of psychological research have given us insights into how the brain learns, and we can apply these to the grieving process. Psychologists have defined learning as "the process by which changes in behavior arise as a result of experiences interacting with the world."[1] Though learning capacity and cognitive functioning cover a wide range of abilities, even within the normal population, we can say, in the broadest terms, that learning improves our ability to adapt. The great thing about learning is that it is a capacity, and we can increase our capacity. Our brain has plasticity we can harness for learning. Psychologist Carol Dweck calls this a growth mindset.[2] We all have different cognitive capacities, but we all still have the opportunity to learn. Those with very little prior knowledge can be exposed to new information, or to grief education. Those who have a grief disorder can be given feedback in psychotherapy about how rumination and avoidance might be affecting their ability to learn. As close friends and family, we can give people who are grieving the opportunity, space, kindness, and encouragement they need to practice new ways of living and realize new insights.

One key to a growth mindset is trying new strategies when we feel stuck, when we feel like we are not learning anything new about our experience of loss. Initially, in acute grief, we are just trying to stay upright, to put one foot in front of the other, and hope that those feet are wearing matching shoes. As time passes, being stuck often feels like we are just going through the motions. Being stuck means we are unable to be creative, or feel love, or help others in our lives. New strategies for learning during this later point in grieving mean having a repertoire, a toolkit of things to try when we feel overwhelmed by pangs of grief or overwhelmed by the new and stressful reality we are living. We can look for these tools from others who have gone before us.

Grieving is as old as human relationships, and that universality connects us to our ancestors and to our present community. Extrapolating from what Dweck writes, if you find yourself saying, "I am not able to adapt to life after loss," try adding "yet" to the end of that sentence. The frustration in learning about your new world, the despair that you will never create a restored life, are feelings created when your brain is growing and changing. Your brain is sorting out what works and what does not. If you feel as though you are treading water, or barely keeping from going under, it is time to try some new approaches to your memories, your emotions, and your relationships. Learning how others have restored a meaningful life can provide new things to try. Your pastor, your grandmother, your favorite novelist or blog writer, a psychologist—consult with someone new, with whom you have not yet talked about your personal experience of grief. Choose someone who has had experience with grieving. Ask what they did to cope, or, even more likely, what they still do to cope. Try out these new approaches, try out what worked for them even if you feel foolish, and then pay attention to what works, what actually makes you feel better in the present moment. Even if none of their ideas work, you may at least feel more connected to someone, more connected to humanity. And since connection is part of what is missing from a grief-stricken life, there is opportunity in it.

Grief 101

I teach what I have learned about grief in a Psychology of Death and Loss class to undergraduates in their junior or senior year. I love teaching this class, and by all accounts, students love taking it.

This may surprise you, since death and loss may not sound like topics a young person might choose to spend sixteen weeks thinking, talking, reading, and writing about. As for me, a student once told me I was "way too happy" to be teaching such a course. Perhaps they expect me to seem depressed or to wear black all the time, and the simple fact that I seem comfortable at the podium talking about death shocks them a little. I do not sugarcoat what I am conveying, and more than once I have teared up while speaking about the death of a child or a genocide, and they probably hear the words *death* and *dying* more during one semester in my class than they have in the rest of their college careers.

But our conversations dig into the real stuff of life, and young people are aching to talk about these things, searching for answers. I look forward to entering the lecture hall of 150 wooden chairs with swing-down writing arms, and I never know quite where the conversation will lead. These college students always surprise me by the amount of life and death they have already experienced. A disturbing number of them have had a friend die by suicide. Many have helped to care for aging relatives and had hospice in their homes. Some of them have worked as grief support volunteers for kids, or trained as emergency medical technicians (EMTs).

We discuss what acute grief looks like, and more than one student has shared that the only time they saw their father cry was following a family member's death. We apply information about a child's cognitive development to understand how grasping the abstract nature of death changes as they grow up. I teach them how to have a conversation with a friend who might be considering suicide, and what to do if they are. Over the Thanksgiving break, they bring home the forms to create a living will with their parents or grandparents, or for themselves, and we practice asking family members what is important to them for their care at the end of life.

After the shootings at a concert in Las Vegas in 2017, a student asked if we could talk about it in class. Then she blurted out from the lecture hall that she was terrified. Several of the students had friends who had attended the concert, and I knew I would need to rip up my lecture for the day. Instead, we talked through their experience, what fear of death looks like for them in the modern world, and how they could manage their terror in part by also focusing on the people who acted with incredible heroism.

One of my favorite discussions is a thought experiment that we do on the last day of class. I bring them the breaking news that new medical science has just created a pill we can take to live forever. Then I ask what would change for them if they were immortal? What would they do differently with their lives? We consider a number of variations—would people still get sick, or would they age in this alternate reality? But these are just details. More important answers are about how it would change their plans. Some of them tell me they would leave college, because they could get a degree anytime. Others tell me that they would get multiple degrees, since they had the time and they have so many interests. A big discussion is around whether they would be more or less likely to have children. Would they want to meet every person in the world, since they have the time? What would it mean for governments, for peace negotiations, for foreign aid?

As the fairly rambunctious conversation winds down, I point out to them the surprising implications. What they do with their lives tightly links with their mortality. The finite nature of our life affects what we do, what we value, how we conduct ourselves.

Although they never explicitly include in their decisions and choices that life is time-limited and its length unknowable, having seen through our thought experiment how a change to that reality affects what they would do, puts into perspective the fact that death impacts us every day. Death adds meaning into life, because life is a limited gift. I close by reading to them a quote from the great Zen Master Dōgen: "Life and death are of supreme importance. Time swiftly passes by and opportunity is lost. Each of us should strive to awaken. Awaken! Take heed, do not squander your life."

Learning Is Not Gained Through Advice

What I teach, however, is not advice about what to do. I also do not think that other people can give advice to someone who is grieving. It may surprise you to hear this from a clinical psychologist—but insight just does not work this way. Other people cannot tell us what grief will feel like to us. In fact, I think advice is exactly what makes grieving people hold at arm's length those who would like to help them. People are experts on their own grief, their own life, their own relationships. As a scientist, I am an expert on grief in general, on average. I can expose people to lots of ways of thinking about grief. I can expose them to scientific evidence showing that although we historically thought grief worked in stages, we now know this is not at all how it works. I can expose them to the concepts that guide psychotherapy for those who have disordered grief, or the common ruminations about loss and grief that people get stuck on. I can show them how grief is like learning and explain what helps or hinders our capacity to learn. As a fellow human being, I can share with them the personal things I have done in moments when I was overwhelmed with grief, or times when I was not at all overwhelmed and felt stigmatized because of that. Much of what psychotherapy does is to give people the opportunity, the courage, and the possibility to experience their emotions, their relationships, and their inner thoughts in a different way than they have before.

I cannot tell anyone how their values and their beliefs feed into what they should do with their life. You are already in your newly restored life, full of love and grief and suffering and wisdom. I can only encourage you to stay in the present and try to learn from what happens day to day, and to learn from what works for you. I believe in your ability to solve your problems and to live a meaningful life after having experienced devastating loss.

What I Have Learned

When I revisit the moments surrounding my mother's death, I am transported back to the old hospital of my hometown. After that terrible flight to Montana that I was loath to take, I went straight there. Thinking of her death in the years that followed, my memories of that hospital room activated painful thoughts about her suffering, her anxiety, her depression, and the guilt-laden conversations we had in the months before she passed away. For the longest time, my thoughts went directly to regret about having not been more patient with her, more understanding. My thoughts were of my shame about not spending enough time with her. But in recent years when I think of her death, I think of walking into that hospital room when I arrived and finding she had slipped into a comatose state. I looked down at her in the hospital bed. Her face, so utterly familiar, was waxy and yellow, a combination of the effects of years of chemotherapy and her liver shutting down. But the remarkable thing, the thing that drew my attention, and the thing to which my thoughts return to now, was that not a single wrinkle troubled her forehead. Her forehead was completely smooth, unlike the characteristic furrowed brow she wore in life, telegraphing her inner turmoil. In her last hours on Earth, she appeared to have found peace. She did not need me in order to find peace in the end.

Coming into contact with death when we lose a loved one can be overwhelming. It can fill us with awe, and it can cause us to reevaluate our view of the world, of our life, of our relationships. Death changes us, and we cannot function in the world in the same way we once did. If you now understand, deeply and truly, that people we love can disappear forever, it changes how we love, what we believe, and what we value. This reevaluation is a form of learning. Coming into contact with great suffering, experiencing the devastation of wanting so desperately for your loved one to be here as they once were, and suffering the reality that this is no longer the case, can be overwhelming. These experiences are part of the nature of getting to be born and getting to live. We will be separated from our loved ones, in ways large and small, from death, to divorce, to misunderstanding, to unintentional slights. Going through these painful events can also bring us together. Once you have experienced deep grieving, you walk through a doorway to a whole community of people that you would otherwise never have understood and empathized with. You probably would not choose this door, if the choice were yours. And yet, here you are on the other side, with knowledge about yourself and a marvelous brain that you can utilize to build and navigate a new world.

Acknowledgments

Having become expert in one field, I recognized I was a complete newbie when it came to publishing, and I am so grateful to quite a few people who both encouraged me and showed me the ropes. First, deep gratitude goes to my agent, Laurie Abkemeier, who took a chance on me and answered endless questions with deep knowledge of the industry and with speediness that was reassuring for an anxious first-time author. She, and everyone at DeFiore and Company, had such insight not only into the issues that an academic-turned-author faces but also into the underlying issues of social justice that influence who and what gets published, which made a great impression on me. To my editor, Shannon Welch, who became passionate about this book and championed it through wildfires and pandemic, I am so grateful for having found an editor who really "got" the point of it and was also willing to give me detailed, thoughtful comments and suggestions. Thanks to my second editor Mickey Maudlin, who swooped in at the final hour when needed, and brought the book across the line. Additionally, Aidan Mahony, Chantal Tom, and the entire team at HarperOne were superbly professional. Thanks to Kent Davis, without whose initial encouragement I would never have submitted the idea for this book to an agent. For Anna Visscher, Andy Steadham, Dave Sbarra, and Saren Seeley, who each read the entirety of the first draft, I am grateful for your time and your kind comments on what worked well, and what needed improvement. For all the academic colleagues who read paragraphs or sections on their own work, I am impressed with your generosity in contributing to the communication of science. Thanks to Tanja at the NOEN café in Utrecht, who provided delicious coffee and tasty sandwiches made on incredible bread, but more importantly, gave me a morning destination for writing and a warm sense of camaraderie when I was a stranger in the Netherlands. To the trivia gang, thank goodness for Thursday nights and Sunday afternoons. To all of my students in the Grief, Loss, and Social Stress (GLASS) lab, the accountability of our writing group enabled me to write in the face of so many other important tasks. Deep thanks to my big sister, Caroline O'Connor, and my best friend, Anna Visscher, who have been there every step of the way through all the events of my life and for being at the end of the phone line any time of night or day. To Jenn, thank you for all the good years. To Rick, my deepest gratitude for following me around the world while I wrote, and for living our less-is-more lives together. To my parents, I am grateful for your endless confidence in me, and for sharing the beautiful process of your life and your death. Finally, to the bereaved people who have shared your stories with me over many years, I admire your persistence in the face of great loss, and your willingness to engage with a scientific process that gives us a lens into your mind, your brain, and your spirit.

Notes

Introduction

1. H. Gündel, M. F. O'Connor, L. Littrell, C. Fort, and R. Lane (2003), "Functional neuroanatomy of grief: An fMRI study," *American Journal of Psychiatry* 160: 1946–53.
2. G. A. Bonanno (2009), *The Other Side of Sadness: What the New Science of Bereavement Tells Us about Life after Loss* (New York: Basic Books).

Chapter 1: Walking in the Dark

1. A. Tsao, M. B. Moser, and E. I. Moser (2013), "Traces of experience in the lateral entorhinal cortex," *Current Biology* 23/5: 399–405.
2. J. O'Keefe and L. Nadel (1978), *The Hippocampus as a Cognitive Map* (New York: Oxford University Press).
3. *Meerkat Manor*, Season one, Discovery Communications, *Animal Planet*,

produced by Oxford Scientific Films for Animal Planet, International Southern Star Entertainment UK PLC, producers Chris Barker and Lucinda Axelsson.

4. J. Bowlby (1982), *Attachment* (2nd ed.), vol. 1: *Attachment and Loss* (New York: Basic Books).

Chapter 2: Searching for Closeness

1. A. Aron, T. McLaughlin-Volpe, D. Mashek, G. Lewandowski, S. C. Wright, and E. N. Aron (2004), "Including others in the self," *European Review of Social Psychology*, 15/1: 101–32, https://doi.org/10.1080/10463280440000008.

2. Y. Trope and N. Liberman (2010), "Construal level theory of psychological distance," *Psychological Review* 117: 440, doi: 10.1037/a0018963.

3. C. Parkinson, S. Liu, and T. Wheatley (2014), "A common cortical metric for spatial, temporal, and social distance," *Journal of Neuroscience* 34/5: 1979–87.

4. R. M. Tavares, A. Mendelsohn, Y. Grossman, C. H. Williams, M. Shapiro, Y. Trope, and D. Schiller (2015), "A map for social navigation in the human brain," *Neuron* 87: 231–43.

5. M. K. Shear (2016), "Grief is a form of love," in R. A. Neimeyer, ed., *Techniques of grief therapy: Assessment and intervention*, 14–18 (Abingdon: Routledge/Taylor & Francis Group).

6. K. L. Collins et al. (2018), "A review of current theories and treatments for phantom limb pain," *Journal of Clinical Investigation* 128/6: 2168.

7. G. Rizzolatti and C. Sinigaglia (2016), "The mirror mechanism: A basic principle of brain function," *Nature Reviews Neuroscience* 17: 757–65.

8. N. A. Harrison, C. E. Wilson, and H. D. Critchley (2007), "Processing of observed pupil size modulates perception of sadnss and predicts empathy," *Emotion* 7/4: 724–29, https://doi.org/10.1037/1528-3542.7.4.724.

Chapter 3: Believing in Magical Thoughts

1. https://www.nytimes.com/2004/02/07/arts/love-that-dare-not-squeak-its-name.html.

2. K. Cronin, E. J. C. van Leeuwen, I. C. Mulenga, and M. D. Bodamer (2011), "Behavioral response of a chimpanzee mother toward her dead infant," *American Journal of Primatology* 73: 415–21.

3. D. Tranel and A. R. Damasio (1993), "The covert learning of affective valence does not require structures in hippocampal system or amygdala," *Journal of Cognitive Neuroscience* 5/1 (Winter): 79–88, https://doi.org/10.1162/jocn.1993.5.1.79.

4. Ibid.

Chapter 4: Adapting Across Time

1. "Elisabeth Kübler-Ross" (2004), *BMJ* 2004; 329:627, doi: https://doi.org/10.1136/bmj.329.7466.627.

2. J. M. Holland and R. A. Neimeyer (2010), "An examination of stage theory of grief among individuals bereaved by natural and violent causes: A meaning-oriented contribution," *Omega* 61/2: 103–20.

Chapter 5: Developing Complications

1. I. R. Galatzer-Levy and G. A. Bonanno (2012), "Beyond normality in the study of bereavement: Heterogeneity in depression outcomes following loss in older adults," *Social Science & Medicine* 74/12: 1987–94.

2. S. Freud (1917), *Mourning and Melancholia*, vol. XIV in The Standard Edition of the Complete Psychological Works of Sigmund Freud (1914–1916): *On the History of the Psycho-Analytic Movement, Papers on Metapsychology and Other Works*, pp. 237–58, https://www.pep-web.org/document.php?id=se.014.0237a.

3. H. G. Prigerson, M. K. Shear, S. C. Jacobs, C. F. Reynolds, P. K. Maciejewski, P. A. Pilkonis, C. M. Wortman, J. B. W. Williams, T. A. Widiger, J. Davidson, E. Frank, D. J. Kupfer, and S. Zisook (1999), "Consensus criteria for traumatic grief: A preliminary empirical test," *British Journal of Psychiatry*, 174: 67–73.

4. H. C. Saavedra Pérez, M. A. Ikram, N. Direk, H. G. Prigerson, R. Freak-Poli, B. F. J. Verhaaren, et al. (2015), "Cognition, structural brain changes and complicated grief: A population-based study," *Psychological Medicine* 45/7: 1389–99, https://doi.org/10.1017/S0033291714002499.

5. H. C. Saavedra Pérez, M. A. Ikram, N. Direk, and H. Tiemeier (2018), "Prolonged grief and cognitive decline: A prospective population-based study in middle-aged and older persons," *American Journal of Geriatric Psychiatry* 26/4: 451–60, https://doi.org/10.1016/j.jagp.2017.12.003.

6. F. Maccallum and R. A. Bryant (2011), "Autobiographical memory following cognitive behaviour therapy for complicated grief," *Journal of Behavior Therapy and Experimental Psychiatry* 42: 26–31.

7. M. K. Shear, Y. Wang, N. Skritskaya, N. Duan, C. Mauro, and A. Ghesquiere (2014), "Treatment of complicated grief in elderly persons: A randomized clinical trial," *JAMA Psychiatry* 71/11: 1287–95, doi:10.1001/jamapsychiatry.2014.1242.

8. P. A. Boelen, J. de Keijser, M. A. van den Hout, and J. van den Bout (2007), "Treatment of complicated grief: A comparison between cognitive-behavioral therapy and supportive counseling," *Journal of Consulting and Clinical Psychology* 75: 277–84.

Chapter 6: Yearning for Your Loved One

1. M. Moscovitch, G. Winocur, and M. Behrmann (1997), "What is special about face recognition? Nineteen experiments on a person with visual object agnosia and dyslexia but normal face recognition," *Journal of Cognitive Neuroscience* 9/5: 555–604.

2. H. Wang, F. Duclot, Y. Liu, Z. Wang, and M. Kabbaj (2013), "Histone deacetylase inhibitors facilitate partner preference formation in female prairie voles," *Nature Neuroscience*, http://dx.doi.org/10.1038/nn.3420.

3. J. Holt-Lunstad, T. B. Smith, and J. B. Layton (2010), "Social relationships and mortality risk: A meta-analytic review," *PLoS Medicine* 7(7): e1000316, doi:10.1371/journal.pmed.1000316.

4. M. F. O'Connor, D. K. Wellisch, A. L. Stanton, N. I. Eisenberger, M. R. Irwin, and M. D. Lieberman (2008), "Craving love? Complicated grief activates brain's reward center," *NeuroImage* 42: 969–72.

5. B. Costa, S. Pini, P. Gabelloni, M. Abelli, L. Lari, A. Cardini, A. Muti, C. Gesi, S. Landi, S. Galderisi, A. Mucci, A. Lucacchini, G. B. Cassano, and C. Martini (2009), "Oxytocin receptor polymorphisms and adult at-

tachment style in patients with depression," *Psychoneuroendocrinology* 34/10 (Nov.): 1506–14, doi: 10.1016/j.psyneuen.2009.05.006.

6. K. Tomizawa, N. Iga, Y. F. Lu, A. Moriwaki, M. Matsushita, S. T. Li, O. Miyamoto, T. Itano, and H. Matsui (2003), "Oxytocin improves long-lasting spatial memory during motherhood through MAP kinase cascade," *Nature Neuroscience* 6/4 (Apr.): 384–90.

Chapter 7: Having the Wisdom to Know the Difference

1. M. F. O'Connor and T. Sussman (2014), "Developing the Yearning in Situations of Loss scale: Convergent and discriminant validity for bereavement, romantic breakup and homesickness," *Death Studies* 38: 450–58, doi: 10.1080/07481187.2013.782928.

2. D. J. Robinaugh, C. Mauro, E. Bui, L. Stone, R. Shah, Y. Wang, N. A. Skritskaya, C. F. Reynolds, S. Zisook, M. F. O'Connor, K. Shear, and N. M. Simon (2016), "Yearning and its measurement in complicated grief," *Journal of Loss and Trauma* 21/5: 410–20, doi: 10.1080/15325024.2015.1110447.

3. D. C. Rubin, M. F. Dennis, and J. C. Beckham (2011), "Autobiographical memory for stressful events: The role of autobiographical memory in posttraumatic stress disorder," *Consciousness and Cognition* 20: 840–56.

4. S. A. Hall, D. C. Rubin, A. Miles, S. W. Davis, E. A. Wing, R. Cabeza, and D. Berntsen (2014), "The neural basis of involuntary episodic memories," *Journal of Cognitive Neuroscience* 26: 2385–99, doi: 10.1162/jocn_a_00633.

5. G. A. Bonanno and D. Keltner (1997), "Facial expressions of emotion and the course of conjugal bereavement," *Journal of Abnormal Psychology* 106/1 (Feb.): 126–37, doi: 10.1037//0021-843x.106.1.126.

6. D. Kahneman and R. H. Thaler (2006), "Utility maximization and experienced utility," *Journal of Economic Perspectives* 20/1: 221–34, doi:10.1257/089533006776526076.

7. P. K. Maciejewski, B. Zhang, S. D. Block, H. G. Prigerson (2007), "An empirical examination of the stage theory of grief," *Journal of the American Medical Association* 297(7): 716–23, erratum in *JAMA* 297/20: 2200, PubMed PMID: 17312291.

Chapter 8: Spending Time in the Past

1. W. Treynor, R. Gonzalez, and S. Nolen-Hoeksema (2003), "Rumination reconsidered: A psychometric analysis," *Cognitive Therapy and Research* 27/3 (June): 247–59.
2. M. C. Eisma, M. S. Stroebe, H. A. W. Schut, J. van den Bout, P. A. Boelen, and W. Stroebe (2014), "Development and psychometric evaluation of the Utrecht Grief Rumination Scale," *Journal of Psychopathology and Behavioral Assessment* 36:165–76, doi: 10.1007/s10862–013–9377-y.
3. M. C. Eisma, H. A. Schut, M. S. Stroebe, P. A. Boelen, J. Bout, and W. Stroebe (2015), "Adaptive and maladaptive rumination after loss: A three-wave longitudinal study," *British Journal of Clinical Psychology* 54:163–80, https://doi.org/10.1111/bjc.12067.
4. M. S. Stroebe et al. (2007), "Ruminative coping as avoidance: A reinterpretation of its function in adjustment to bereavement," *European Archives of Psychiatry and Clinical Neuroscience* 257: 462–72, doi: 10.1007/s00406–007–0746-y.
5. M. C. Eisma, M. Rinck, M. S. Stroebe, H. A. Schut, P. A. Boelen, W. Stroebe, and J. van den Bout (2015), "Rumination and implicit avoidance following bereavement: an approach avoidance task investigation," *Journal of Behavior Therapy and Experimental Psychiatry* 47 (Jun): 84–91, doi: 10.1016/j.jbtep.2014.11.010.
6. M. C. Eisma, H. A. W. Schut, M. S. Stroebe, J. van den Bout, W. Stroebe, and P. A. Boelen (2014), "Is rumination after bereavement linked with loss avoidance? Evidence from eye-tracking," *PLoS One* 9, e104980, http://dx.doi.org/10.1371/journal.pone.0104980.
7. A. J. Rose, W. Carlson, and E. M. Waller (2007): "Prospective associations of co-rumination with friendship and emotional adjustment: Considering the socioemotional trade-offs of co-rumination," *Developmental Psychology* 43/4: 1019–31, doi: 10.1037/0012–1649.43.4.1019.

Chapter 9: Being in the Present

1. J. Warner, C. Metcalfe, and M. King (2001), "Evaluating the use of benzodiazepines following recent bereavement," *British Journal of Psychiatry* 178/1: 36–41.

2. J. M. Cook, T. Biyanova, and R. Marshall (2007), "Medicating grief with benzodiazepines: Physician and patient perspectives," *Archives of Internal Medicine* 167/18 (Oct. 8), doi:10.1001/archinte.167.18.2006.

3. W. W. Seeley, V. Menon, A. F. Schatzberg, J. Keller, G. H. Glover, H. Kenna, et al. (2007), "Dissociable intrinsic connectivity networks for salience processing and executive control, *Journal of Neuroscience* 27: 2349–56.

4. J. D. Creswell, A. A. Taren, E. K. Lindsay, C. M. Greco, P. J. Gianaros, A. Fairgrieve, A. L. Marsland et al. (2016), "Alterations in resting-state functional connectivity link mindfulness meditation with reduced interleukin-6: A randomized controlled trial," *Biological Psychiatry* 80: 53–61, http://dx.doi.org/10.1016/j.biopsych.2016.01.008.

Chapter 11: Teaching What You Have Learned

1. S. J. E. Bruijniks, R. J. DeRubeis, S. D. Hollon, and M. J. H. Huibers (2019), "The potential role of learning capacity in cognitive behavior therapy for depression: A systematic review of the evidence and future directions for improving therapeutic learning," *Clinical Psychological Science* 7/4: 668–92, https://doi.org/10.1177/2167702619830391.

2. C. S. Dweck (2006), *Mindset* (New York: Random House).

Index